Florida's Best Fruiting Plants

Charles R. Boning
Illustrated and Photographed by the Author

Pineapple Press, Inc.
Sarasota, Florida

Inquiries should be addressed to:

Pineapple Press, Inc.
P.O. Box 3889
Sarasota, Florida 34230
www.pineapplepress.com

Library of Congress Cataloging-in-Publication Data

Boning, Charles R.
  Florida's best fruiting plants / Charles R. Boning.
       p. cm.
  Includes bibliographical references and index.
  ISBN-13: 978-1-56164-372-1 (pbk. : alk. paper)
  ISBN-10: 1-56164-372-6 (pbk. : alk. paper)
  1. Fruit-culture—Florida. 2. Fruit—Florida. I. Title.
  SB355.B66 2006
  634.09759—dc22   2006005433

First Edition
10 9 8 7 6 5 4 3 2 1

Design by Charles R. Boning and Shé Heaton
Printed in China

Preface

Florida is experiencing a surge of interest in growing fruit for the home. Dooryard orchards have sprung up across the state. Tree sales, such as those sponsored by the Rare Fruit Council International, attract huge crowds. Fruit festivals draw thousands. Even home-supply warehouses have capitalized upon the growing popularity of rare and unusual fruit by stocking cultivars of lychee, persimmon, and passionfruit.

Much of this enthusiasm can be traced to waves of immigration from Latin America and Asia, where many of the best species described within these pages originated. However, the attraction transcends geographic boundaries and regional differences. Those who migrate to Florida from northern states yearn for something different—a taste of the tropics. Meanwhile, many long-term residents are only beginning to discover the bounty available from their own backyards. The opportunity to grow fruit that is uncommon or prohibitively expensive holds special appeal.

While several books have addressed the cultivation of subtropical fruit, most lack regional focus. This book has been designed from the outset as a reference for Florida homeowners and Florida growers. It provides detailed plant profiles, replete with illustrations, distribution maps, and photographs. Each profile furnishes up-to-date information regarding cultivars, climate requirements, cultivation, and crop management. We at Pineapple Press are confident that this guide will become an indispensable reference for homeowners, nurseries, landscapers, and commercial growers.

Contents

Preface   2

Introduction   4

How to Use This Book   6

Introduction

The lore of fruit permeates Florida's culture and history. Even before the 1500s, when the Spanish planted Florida's first citrus tree, indigenous people enjoyed the fruit borne by native plants. Over the last century and a half, Florida has become a hub of research, production, and distribution of many fruit crops. The state's diverse climate has made it a repository for the world's most valuable fruiting species.

This book is the first comprehensive guide to Florida's fruit-bearing plants. It profiles varieties that are suitable for planting in every region of the state—from the temperate panhandle to the near-tropical Keys. It discusses the best exotic species, such as the mango, and the best native species, such as the pawpaw. It includes familiar plants, such as the orange, and dozens of rare and obscure plants, such as the Ogeechee lime and blue grape.

PURPOSE The primary purpose of this book is to acquaint Floridians with the many options available for dooryard planting. The fruit borne by each species described within these pages embodies a unique mix of flavors, textures, aromas, colors, and shapes. The diversity and abundance of Florida's fruit crops is remarkable. Indeed, because its climate ranges from near tropical to temperate, Florida likely surpasses every other state in terms of the number of fruit species that will grow within its borders.

Most residents have never sampled the aromatic pulp of the jaboticaba, the custardlike flesh of the pawpaw, or the crisp, sweet arils of the jackfruit. It is a pity that so few are familiar with these fruits. In flavor and appeal, they rival or surpass the very best varieties found on produce shelves. Consequently, this book promotes several uncommon fruits that deserve wider recognition and distribution.

In addition, this book presents candid assessments of the attributes and defects of each variety. It details climate requirements, soil preferences, and maintenance requirements. Wherever possible, this book describes superior cultivars within a given species. Thus, in every respect, this book has been crafted to assist the gardener in making informed planting decisions.

Another important goal of this book is to encourage home fruit production and small-scale agriculture. Florida's population rose from under 3 million in 1950, to 17 million in 2005. Officials predict that the population will swell to 20 million by 2025. As land prices soar, fruit groves and family farms are routinely bulldozed to make way for residential lots and strip malls. Every day, fragments of Florida's agricultural heritage are lost.

The Redlands Agricultural District of Miami-Dade County is a poignant example of this trend. The Redlands is a pioneering community specializing in tropical agriculture. It is unique, the only large neighborhood of its type in the continental United States, composed of a patchwork of family farms and nurseries. Even today, it contains thousands of acres of fruit groves. Tractors still rumble down the streets. Roadside stands still display colorful mounds of lychees, guavas, and mangos. But the shadow of suburban sprawl has cast itself upon this landscape. The future of the Redlands—like that of many other farming communities throughout Florida—is uncertain. As agricultural land becomes scarce, the impetus for dooryard production increases. Ultimately, the key to preserving the state's tradition of small-scale agriculture may lie with the individual homeowner.

SELECTION AND SCOPE This book does not cover every fruit-bearing plant that will grow in Florida. It focuses instead on 80 plants of merit. The species included are those that have been successfully cultivated in Florida, those that hold special appeal for landscaping, and those that are most likely to bear regular crops of high quality fruit in a dooryard setting. Collectively, they represent a cross section of the world's noteworthy families of fruiting plants.

Special effort has been made to include examples of native fruit trees. This book profiles over a dozen plants that are wholly or partly indigenous. The fruit borne by some of these plants does not measure up to that borne by the best exotic species. Nevertheless, as a result of their toughness and beauty, these natives make outstanding additions to the home landscape. A few native species, such as the pawpaw, muscadine grape, and red mulberry, bear fruit of world-class quality.

Citrus has long been the state's most important fruit crop. In keeping with this observation, this book profiles five members of the *Citrus* genus and several relatives. However, this book does not focus on citrus to the exclusion of other types of fruit. To some extent, the low price and widespread availability of citrus lessens its appeal as a fruit for the home garden. In addition, the spread of diseases such as citrus canker and citrus greening casts some doubt on the future of citrus as a dooryard crop.

Consequently, this book presents an array of alternatives to citrus. Many of the varieties profiled within these pages exceed citrus in one or more respects—in taste, disease resistance, or other qualities. Some are extremely hardy, producing abundant crops even when deprived of basic care. Some bear multiple crops, thus assuring a year-round harvest. Citrus is an extraordinary crop. However, it is not the only worthwhile fruit that will grow in Florida.

WHY GROW FRUIT? No hobby offers greater satisfaction than raising fruit in the home garden. It is an inexpensive pursuit that can be enjoyed by anyone whose residence is surrounded by a small patch of green. It eventually pays for itself through reduced grocery bills, nutritional benefits, and added landscape value.

Even an inexperienced gardener can succeed at growing an amazing array of species. Many of the plants described in this book are resilient and will quickly rebound from any harm inflicted by novice error. Raising fruit trees is a process of discovery and adjustment. All that is required is a spirit of adventure and a small commitment of time and effort.

Fruit harvested directly from the tree is almost always superior in taste to that purchased from the supermarket. While most Floridians have eaten store-bought mangos, few have sampled the fruit at its best. Most supermarkets peddle imported, second-rate varieties, picked prematurely and treated to stave off decay. These bear little resemblance to fresh Florida mangos. Premium cultivars, harvested when ripe, have a unique, perfumed smell, and sweet, rich, melting flesh.

Florida residents have the ability to grow fruit that is not available in stores at any price. They can choose fruit based on quality and taste, ignoring those marketing concerns that guide the decisions of produce managers. The homeowner thus operates in a world unfettered by considerations such as long shelf life, ability to endure rough handling, low price, and attractive coloration. When making planting decisions, the homeowner has the ability to select the very best from among hundreds of species and thousands of cultivars.

Harvesting and eating the fruit is not the only pleasure that awaits those who plant a fruit garden. Time spent among the plants is a superb form of relaxation and renewal. In the garden, one can observe the comings and goings of birds and other wildlife, flowering and fruiting cycles, the change of seasons, and other natural rituals. Many of the plants described in this book double as attractive ornamentals, with lush foliage and showy flowers. The homeowner can derive considerable satisfaction from arranging and displaying these plants within the landscape. Perhaps, though, the most rewarding aspect of fruit gardening is the ability to share the experience with others—to introduce friends and acquaintances to unfamiliar plants and delicious new flavors. It is in this spirit that Pineapple Press introduces *Florida's Best Fruiting Plants.*

Even the ultra-tropical breadfruit (left) and mangosteen (right) will fruit in the Keys, showing the diversity of Florida's climate.

How to Use This Book

This book is divided into two parts. Part I presents basic information related to cultivation. It contains a discussion of Florida's climate, growing conditions, pests, diseases, and other topics. It walks the gardener through the process of raising fruit trees. It also warns of several mistakes to avoid. Part II—the heart of this book—contains profiles of 80 fruiting plants. These are arranged alphabetically based on the common name most often used in Florida.

FEATURES OF THE FRUIT PROFILES The fruit profiles furnish detailed information in a simple, consistent format. Each profile is two or four pages long. The first page relays basic data, including the plant's common name, scientific name, and classification. The most prominent feature is a full-color **illustration** depicting the fruit and leaves of a typical plant. A **fruiting calendar**, presented as a horizontal bar, shows the months during which the plant is likely to bear fruit. Red indicates peak production, orange indicates moderate or occasional production, and yellow indicates low or infrequent production.

On the top right-hand side of the page, a state **map** shows the plant's range. Portions shaded in dark green denote areas to which the plant is best adapted. Portions shaded in light green show areas where the plant will grow, but where growing conditions are not optimal. Portions shaded in yellow show areas to which the plant is only marginally adapted. Beneath the range map, a plant **silhouette** shows the average dimensions and growth habit of a mature tree. The lower right side of the first page of each profile contains a **characteristics chart** that ranks various attributes from one to five stars. A text box containing a brief description of common hazards appears below the characteristics chart. Readers should note that plants without any widely known **hazards** may nevertheless produce adverse reactions in some individuals. Therefore, any unfamiliar fruit should be sampled cautiously and in small quantity. Medical attention should be sought at the first sign of distress. The reader should presume that all parts of the plant, other than the fruit, are poisonous.

The text follows a standard order and, in most cases, is organized under the following subheadings: Geographic Distribution, Tree Description, Fruit Characteristics, Cultivars, Relatives, Climate, Cultivation, and Harvest and Use. Cultivars within a species are ordered in terms of their suitability for dooryard planting in Florida. All rankings and ratings within this book represent the subjective opinion of the author.

TERMINOLOGY While this book contains a glossary, some often-used terms require a preliminary explanation. A plant's scientific name is composed of two parts. The first identifies the genus. The second differentiates the organism from others within the genus. In addition to the scientific name, most species have one or more common names. Occasionally, a single common name encompasses several species.

Within this book, the term "variety" is used in a general sense and embraces species, cultivars, strains, seedling races, and other variants. The term 'cultivar' is short for *culti*vated *vari*ety. When used in this book, it refers strictly to plants that have been cloned to preserve the genetic characteristics of the parent. Some species, such as the mango, have given rise to hundreds of named cultivars.

THE CLASSIFICATION OF FRUITING PLANTS Plants are grouped and classified through an array of features. Understanding the relationship between species can aid the gardener in several ways. Related species often require similar regimens of care, succeed under similar conditions, and suffer from identical problems. Knowing that a species belongs to a particular family can assist the grower when making planting decisions.

A few of the plants described within these pages are isolated species within families that do not contain other fruit-bearing members. However, most belong to one of a dozen fruit-rich families: Anacardiaceae (Spondia), Annonaceae (Annona), Arecaceae (Palm), Clusiaceae (Garcinia), Cucurbitaceae (Gourd), Moraceae (Mulberry), Myrtaceae (Myrtle), Passiflora (Passionflower), Rosaceae (Rose), Rutaceae (Citrus), Sapindaceae (Soapberry), and Sapotaceae (Sapote). Each plant profile identifies the family to which the plant belongs and lists related fruiting species.

I

The Basics of Growing Fruit in Florida

Raising fruit trees entails a combination of science and art. Scientific techniques of fertilization, pest control, and irrigation play a vital role in production. The art of raising fruit trees involves selecting varieties, making aesthetic choices within the landscape, and interpreting conditions within the garden based on insights gained over time.

The process is relatively simple. First, the grower must understand local conditions and select species that will succeed under those conditions. Second, the grower must plant the specimen in a manner designed to ensure its long-term survival. Third, the grower must meet the needs of the plant by applying the proper regimen of care. Finally, the grower must overcome challenges posed by pests, diseases, and weather.

FLORIDA GROWING CONDITIONS

Florida is composed of many regions with diverse climates, soils, and growing conditions. This variability gives rise to unmatched opportunities. However, it also provides growers with unique challenges.

GEOGRAPHY Florida is located between 24°30' and 31°00' north latitude. The peninsula—an extension of the coastal plain—is the state's most prominent geographic feature. It runs south for 400 miles, dividing the Atlantic Ocean from the Gulf of Mexico. Florida's terrain is remarkably flat. The central panhandle contains a series of low hills that comprise the southern foothills of the Appalachian Mountains. Another section of uplands, known as the Central Ridge, extends south along the spine of the peninsula. The highest point in the state is a mere 345 feet above sea level. Nevertheless, altitude can have a significant impact on agriculture. Trees planted on high or sloping land are less susceptible to freeze damage than those planted in low areas, where cold air tends to pool.

Most of the state sits atop a porous limestone foundation. Several large aquifers permeate the limestone, giving rise to sinkholes, springs, and other features. In addition to this vast underground water supply, Florida has an estimated 7,800 lakes, with the heaviest concentration occurring in the central part of the state. This surface water can have a moderating influence on temperature.

SOIL The soil covering most of the peninsula consists of infertile sand. A clay hardpan often lies several inches or several feet beneath the surface. The southeast coast features outcrops of oolitic limestone. Other limestone outcrops are present on the lower Gulf coast and along the Suwannee River valley. Red clay is prevalent in the uplands of north Florida.

Heavy machinery may be required to dig a planting hole in the oolitic limestone soil of Miami-Dade County.

The pH scale, which runs from 0 to 14, measures the hydrogen ion concentration. Acidic soils have a pH value of less than 7.0. Alkaline soils have a pH value of greater than 7.0. Most fruit trees prefer neutral or mildly acidic soil. Some, such as the blueberry, blackberry, miracle fruit, and pineapple, demand acidic soil. Few fruiting plants, apart from the guava, jujube, olive, and prickly pear, will tolerate severe alkalinity. However, with proper care, many species will make acceptable growth in soils that are mildly alkaline.

The miracle fruit requires acidic soil. The fruit is about the size of a jelly bean.

CLIMATE Florida's climate provides residents with the opportunity to grow an incredible variety of fruit-bearing plants. North Florida lies within the temperate zone. Winter

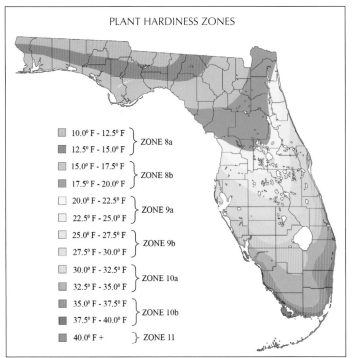

PLANT HARDINESS ZONES

10.0° F - 12.5° F	ZONE 8a
12.5° F - 15.0° F	
15.0° F - 17.5° F	ZONE 8b
17.5° F - 20.0° F	
20.0° F - 22.5° F	ZONE 9a
22.5° F - 25.0° F	
25.0° F - 27.5° F	ZONE 9b
27.5° F - 30.0° F	
30.0° F - 32.5° F	ZONE 10a
32.5° F - 35.0° F	
35.0° F - 37.5° F	ZONE 10b
37.5° F - 40.0° F	
40.0° F +	ZONE 11

This map is derived from NOAA data and the Plant Hardiness Zone Map published by the United States Department of Agriculture. It shows the average lowest temperature that can be expected on an annual basis.

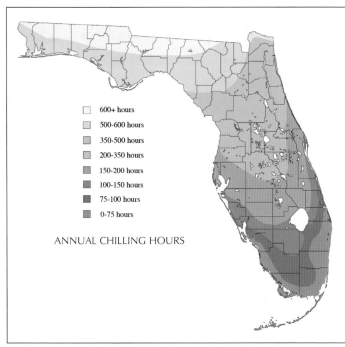

	600+ hours
	500-600 hours
	350-500 hours
	200-350 hours
	150-200 hours
	100-150 hours
	75-100 hours
	0-75 hours

ANNUAL CHILLING HOURS

This map is derived from NOAA data and the Chill Data Map created by the Florida Cooperative Extension Service, Institute of Food and Agricultural Sciences, University of Florida. It shows the approximate number of hours that the temperature remains below 45° F during the average winter.

freezes are frequent and can be severe. During the great cold wave of 1899, Tallahassee experienced a low of −2°F and ice was observed in the Gulf of Mexico. At the same time, summers in north Florida are often warmer than those in south Florida. The state record high temperature of 109°F occurred in Monticello, just a few miles south of the Georgia border.

Much of peninsular Florida lies within the transitional zone between warm temperate and subtropical climates. The weather is characterized by long, humid summers and mild winters. The subtropics occur south of a line beginning near Cape Canaveral, sweeping south to the eastern shore of Lake Okeechobee, then curving northwest to Sarasota. The climate of the Keys is usually classified as near tropical. Key West lies about 100 miles north of the true tropics.

While Florida winters are mild, temperature is still a critical factor limiting the production of subtropical fruit crops. No part of Florida, save for the lower Keys, is immune to the threat of freezing weather. Several years may pass in which freezing temperatures do not occur south of Orlando. Gardeners forget the devastating effect of the last cold snap. They begin to plant subtropical trees in areas far north of the true subtropics. The bonanza never lasts.

A blast of arctic air eventually sweeps down from Canada. It withers foliage, destroys years of growth, and sometimes kills established trees outright. As this cold air travels down the peninsula, it mixes with, and is moderated by, warmer air from the Gulf of Mexico and the Atlantic Ocean. By the time the typical arctic blast arrives in Palm Beach or Fort Myers it has lost most of its punch. This cold air may be capable of generating an overnight frost, but it is rarely capable of killing mature subtropical trees. Yet, even south

8

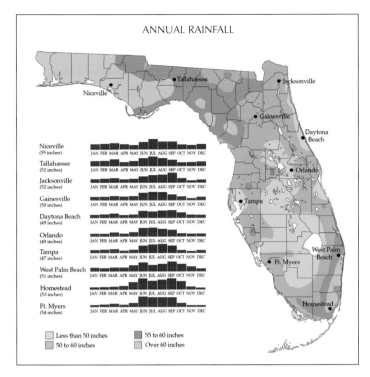

ANNUAL RAINFALL

Niceville (55 inches)
JAN FEB MAR APR MAY JUN JUL AUG SEP OCT NOV DEC

Tallahassee (52 inches)
JAN FEB MAR APR MAY JUN JUL AUG SEP OCT NOV DEC

Jacksonville (52 inches)
JAN FEB MAR APR MAY JUN JUL AUG SEP OCT NOV DEC

Gainesville (50 inches)
JAN FEB MAR APR MAY JUN JUL AUG SEP OCT NOV DEC

Daytona Beach (49 inches)
JAN FEB MAR APR MAY JUN JUL AUG SEP OCT NOV DEC

Orlando (49 inches)
JAN FEB MAR APR MAY JUN JUL AUG SEP OCT NOV DEC

Tampa (47 inches)
JAN FEB MAR APR MAY JUN JUL AUG SEP OCT NOV DEC

West Palm Beach (51 inches)
JAN FEB MAR APR MAY JUN JUL AUG SEP OCT NOV DEC

Homestead (53 inches)
JAN FEB MAR APR MAY JUN JUL AUG SEP OCT NOV DEC

Ft. Myers (54 inches)
JAN FEB MAR APR MAY JUN JUL AUG SEP OCT NOV DEC

Less than 50 inches
50 to 60 inches
55 to 60 inches
Over 60 inches

This map is derived from NOAA data and the Average Annual Precipitation map created by the Spatial Climate Analysis Service of Oregon State University. It shows the average annual rainfall in inches for various cities and locations.

Florida can experience freezing temperatures. Disastrous freezes took place in 1835, 1886, 1894, 1895, 1899, 1901, 1905, 1906, 1917, 1934, 1940, 1951, 1957, 1962, 1977, 1983, 1985, and 1989. The main threat from freezing weather in south Florida occurs in January and February.

Despite these periodic intrusions of cold air, Florida's climate appears to be growing warmer. From the late 1800s through the 1980s serious freezes affected the peninsula every two to four years on average. The 1990s and 2000s, by contrast, have seen unusually warm temperatures and few noteworthy freezes. While some attribute this trend to global warming, others suggest that it is part of a long-term climatic cycle. Because the cause is uncertain, prudence dictates against planting tender plants in vulnerable locations.

Florida's annual precipitation averages 54 inches, making it one of the wettest states in the nation. The Florida peninsula is dominated by a tropical savannah weather pattern. The dry season runs from October through April. The wet season runs from May through September. Because new growth and fruit set often occur in the late winter and spring—before the advent of the rainy season—plants often require supplemental irrigation during this period.

SELECTING A TREE

When purchasing a fruit tree, several considerations come into play. Choosing the wrong tree can transform what might have been years of bounty and enjoyment into years of headaches and scarce harvests. To reduce maintenance requirements, the gardener should select species that are reasonably well adapted to the location.

CULTIVARS Failing to differentiate between cultivars is a common mistake of novice gardeners. Those who think that a mango is a mango may be disappointed when they find they have planted a straggly tree that shades out the yard and bears a scattering of fibrous fruit. It pays to research the choices available. Sampling the fruit of various cultivars is instrumental in making an informed planting decision.

Commercial cultivars are not always well suited to the backyard orchard. Commercial growers are willing to sacrifice flavor for storage and marketing considerations. They favor varieties that ripen over a short period to facilitate harvest and reduce labor costs. Home gardeners, on the other hand, prefer varieties that ripen over a lengthy period so that they can consume the fruit as it comes ripe.

PURCHASE CONSIDERATIONS The selection of an individual tree involves additional choices. Large specimens have a lower mortality rate than small specimens and may bear fruit immediately. At the same time, they are more expensive and may be slow to acclimate to new surroundings. Very young trees, on the other hand, may take years to come into bearing. They are susceptible to cold injury, insect damage, drought, and disease. Specimens in seven- to fifteen-gallon containers represent the best option for most gardeners. Such trees generally bear within a year or two after they are set out.

Most nurseries price plants by container size. Once the gardener has decided on a container size, it makes sense to purchase the largest and most vigorous tree available within the size category. The

9

purchaser should carefully inspect the tree from every angle. Trees can be sorted into a rough pecking order based on the following characteristics: (1) thick trunk diameter, (2) the absence of disease or dead wood, (3) a well-developed (but not container-bound) root system, and (4) healthy foliage arrayed in a manageable shape. Preliminary indications of flowering or fruit production are relatively meaningless on a grafted tree.

OVERCOMING SPACE CONSTRAINTS Some homeowners are content to plant one or two favorite fruit trees. Others are determined to plant as many fruit trees as can be squeezed into a given space. With planning and careful management, the average suburban yard can accommodate a surprising number of trees. The homeowner can space trees more closely than would be permissible in the commercial grove. However, this approach requires aggressive pruning practices. The homeowner can also graft multiple cultivars onto a single rootstock. Finally, those with smaller yards should consider planting dwarf or compact cultivars rather than full-size trees.

PLANTING A TREE

Planting a fruit tree is a simple endeavor. It requires nothing more than digging a hole, placing the root mass within the hole, and filling any voids. Nevertheless, many trees fail to prosper because gardeners violate one or more of the four guidelines set forth below.

PLANT HIGH Perhaps the most common mistake made by beginning gardeners is to plant the tree too deep. Burying the root crown will weaken a tree over time. Trees should generally be planted with the crown of the root system raised 2 or 3 inches above the soil surface. It is far better to place the tree in a hole that is too shallow than in a hole

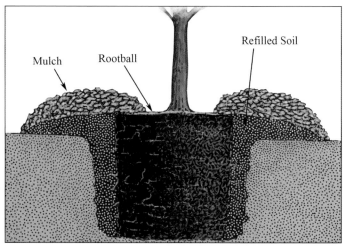

Subtropical and tropical trees should be planted with the top of the root ball raised 2 or 3 inches above the surface of the soil.

that is too deep. Most trees settle slightly within a year after they are planted. By planting high, the gardener can ensure proper drainage, stimulate development, and prevent disease.

AVOID SOIL ADDITIVES Florida's soil is mediocre to poor. As a result, many gardeners go to extensive lengths to prepare and enrich the planting hole. They dig a hole twice as wide and half again as deep as the root system of the plant. After planting, they fill the hole with an organic mix, usually some combination of topsoil, manure, peat, and compost. These preparations are unnecessary and are often harmful. The enriched planting hole discourages roots from extending out into the existing soil, limiting growth and making the tree prone to uprooting in high winds. Manure and other additives can burn tender new roots. In addition, buried organic matter can hold excessive moisture, causing fungal infections and other problems. With few exceptions, the best practice is to plant the tree directly in the soil that is present at the planting site.

DELAY FERTILIZATION In an attempt to stimulate new growth, some gardeners fertilize a new tree immediately after it is planted. This practice is usually harmful. A tree needs time to adjust to new surroundings. Tender feeder roots should be given a chance to develop and extend into nearby soil. To avoid sending the tree into shock, it is best to delay fertilization for about two months after planting. At that point, a light application of slow-release fertilizer may be administered. Some exceptions to this rule exist. For example, the mango seems to benefit from a light fertilization at planting. But, in most cases, immediate fertilization has a detrimental effect on plant development.

WATER REGULARLY Although established trees can endure some drought, regular irrigation is critical for the first four to six months after a tree is planted. A newly planted tree should receive a deep watering twice a week. Root-bound specimens should be watered three or four times a week for two months, and should be watered twice a week thereafter until they are well established. While the construction of a water-holding basin around a tree may be helpful, it is unnecessary where a thick bed of water-holding mulch is present.

CULTIVATION

Some fruit trees live long, productive lives without human intervention. However, as a rule, those that receive proper care tend to be more productive than those that are neglected. The discussion that follows touches on basic cultivation techniques.

IRRIGATION Most fruit trees require irrigation, particularly during fruit set and periods of low precipitation. Plants constantly lose water through evaporation. At the same time they require water to transport nutrients and distribute sugars. If deprived of sufficient water, one of the first functions that a tree will forgo is fruit production. By the time visual symptoms of stress appear—such as wilting or leaf drop—it is likely that serious damage has occurred. The goal is to apply the correct amount of water to supply the needs of the tree.

The average fruit tree benefits from a deep weekly or bi-weekly watering. However, requirements vary with the species, the age of the plant, drainage characteristics of the soil, the season, and the amount of rainfall. The frequency of irrigation should decrease as the tree matures. Irrigation may be unnecessary during the rainy season. During early winter, when growth slows or stops, irrigation can be reduced or discontinued. However, from late February through April—a period marked by dry conditions in peninsular Florida—many species put out new flushes of growth, flower, and set fruit. Irrigation is critical to ensure that these functions take place without interruption.

Watering with a garden hose is an efficient form of irrigation on small lots with young trees. However, as trees mature, they require greater quantities of water spread over a larger area. A minor task can thus evolve into a several-hour or daylong chore. The use of mobile sprinklers can reduce the burden. A simple plastic donut with holes punched at different angles does a remarkably effective job of providing water to the root zone of a single tree. Soaker hoses work well for hedges or planting rows, but tend to deteriorate over time.

Drip irrigation systems supply frequent applications of water at a low flow rate and at low pressure. Such systems are very efficient. However, because they only supply water to a small area, they tend to concentrate root development in areas immediately beneath the supply points. Under-canopy micro sprinklers are slightly less efficient. However, they can be adjusted to cover a broader area. With both systems, the emitters have a tendency to clog. Overhead sprinklers are inefficient. Significant quantities of water are lost to evaporation and run off. Energy use is also high. However, when cold weather threatens, overhead sprinklers can dramatically raise the temperature in the garden.

DRAINAGE Excessive water can be just as harmful as insufficient water. Over-frequent irrigation can encourage development of a shallow root system. A shallow root system can ruin a lawn and may cause the tree to topple in high winds. Few fruit trees will tolerate "wet feet" or saturated soil. Serious problems arise in low areas subject to periodic flooding. Where drainage is inadequate, trees can be planted atop mounds or in mounded rows. Unless the trees are elevated sufficiently, the roots will eventually grow down into wet soil and be visited by root rot and other problems. In areas prone to flooding and in areas with a high water table, the only alternative to mounding is to plant species that will withstand occasional flooding, such as the guava, jaboticaba, mayhaw, Ogeechee lime, pawpaw, and pond apple.

FERTILIZATION Most fruit trees will grow faster and produce greater quantities of fruit if they are fertilized on a regular basis. Slow-release granular fertilizers are ideal for most purposes. They provide even coverage and supply nutrients to the roots over a lengthy period. Fertilizer spikes provide uneven results because nutrients do not leach out horizontally. Nearby plant roots can be burned while distant roots receive no benefit. Inexpensive, soluble fertilizers wash quickly through the soil. They provide no long-term advantage and can damage sensitive plants.

Fertilizer should be evenly distributed beneath the canopy and should cover an area extending past the drip line. Most trees benefit from three or four applications annually. Applications in late February, April, June, and September will ensure maximum growth. Many growers discontinue fertilization from September through February so as not to stimulate tender new growth during cold weather. Many gardeners presume that the best time to apply fertilizer is immediately before it rains. They are incorrect. A heavy rain can wash away nutrients and deposit them below the root zone. The best time to apply fertilizer is immediately after a heavy rain. As newly fallen rainwater filters down through the soil, it pulls nutrients into the root zone through capillary action.

Besides hydrogen, carbon, and oxygen, which are present in the environment, plants require 13 minerals. Nitrogen (N), phosphorus (P), and potassium (K) are the primary macronutrients and are the main ingredients in most fertilizers. Nitrogen is essential for vegetative growth and photosynthesis. It helps produce the rich green color of healthy foliage. Nitrogen deficiency causes a general yellowing of leaves. Symptoms appear first in older leaves. Phosphorus promotes flowering and root growth. It is also essential to photosynthesis. Plants suffering from phosphorus deficiency are often stunted. Although the foliage may remain green, leaves tend to be small, older leaves are shed prematurely, and scorching may occur on leaf tips and margins. Potassium assists the plant in building proteins and is important to fruit production. Symptoms of potassium deficiency appear as a pale mottling or as a browning or curling of leaf margins. Calcium (Ca), magnesium (Mg), and sulfur (S), are also classified as macronutrients, but are usually present in sufficient quantity to support growth. Micronutrients essential for plant growth include boron (B), copper (Cu), iron (Fe), chlorine (Cl), manganese (Mn), molybdenum (Mo), and zinc (Zn).

While requirements vary among species, most fruit trees will achieve optimal growth if they receive regular applications of a quality palm fertilizer, such as a 10–3–10 formula with micronutrients. A balanced 10–10–10 formula may be applied to mature trees during periods of flowering and fruit development. A high-analysis, six-month, time-release fertilizer will serve to accelerate the growth of young specimens.

WEED SUPPRESSION A weed is any plant that grows in a location where it is not wanted. Grasses become weeds when they encroach on areas beneath the canopy of a fruit tree. They retard the tree's growth by competing for nutrients and moisture. Most gardeners use a combination of mulch and herbicides to suppress weeds around the base of a tree. Herbicides are chemical compounds that interfere with plant growth or metabolism. The herbicide Glyphosate is widely used in commercial and residential applications. It is highly effective, has relatively low toxicity, and is not absorbed by the bark or roots of established trees.

Several precautionary notes are in order. When applying an herbicide, the gardener must not allow the spray to touch the leaves of the tree. One moment of inattention can result in disaster. The task should never be performed under breezy conditions. Weed-and-feed lawn fertilizers should not be used in the vicinity of fruit trees. Some contain powerful herbicides that leach into the soil and that are taken up by roots that extend under grassy areas. These toxins can weaken or kill the tree over time. Finally, weed whackers should never be used around the base of fruit trees. They destroy bark and can girdle and kill the tree.

MULCH Organic mulch serves at least six important purposes. First, it conserves soil moisture by limiting evaporation and runoff. Second, it adds organic matter and nutrients to the soil. Third, it draws earthworms into the root zone and encourages the development of beneficial microbes and mycorrhizal fungi.

Fourth, it reduces the population of nematodes, microscopic worms that can harm the roots of the plant. Fifth, it moderates the temperature of the top layer of soil. Sixth, it suppresses weed growth.

Fruit trees surrounded by a thick layer of mulch tend to be far more vigorous than those growing from a patch of bare earth. Mulch should be distributed in a ring beginning 6 inches away from the trunk and ending just beyond the drip line formed by the canopy. Mulch that accumulates around the trunk can lead to the transfer of fungal rot and disease. The mulch bed should be maintained at a thickness of 6 to 8 inches.

Some gardeners place a layer of weed block fabric beneath the mulch. This step is unnecessary. The material tends to deteriorate over time and is unsightly when exposed. It may also slow the migration of soil-based organisms into the mulch and may inhibit the dispersal of organic matter into the soil. Grass clippings should not be

The jaboticaba, like many subtropical fruit trees, thrives when provided with a thick layer of mulch.

used as mulch. They break down rapidly and may harbor numerous weed seeds.

PRUNING Some fruit trees require pruning to develop proper shape, to remove crossing or damaged limbs, and to control size. The overall objective is to develop a strong framework of limbs that will support a heavy fruit crop. Pruning cuts are of two types. A heading cut removes shoot terminals or branch tips. It encourages the growth of buds beneath the cut, leading to thicker, bushier growth. A thinning cut removes an entire shoot back to a main branch or side shoot. It is used to open up trees and to promote light penetration. Tropical and subtropical species are usually pruned after harvest, in the late summer or early fall. Deciduous species are usually pruned during periods of dormancy.

SUN AND SHADE All fruiting plants require light to conduct photosynthesis. Most prosper in full sun. As a general rule, the greater the exposure to sunlight, the greater the production of fruit. However, young plants of certain species, air layers, and newly grafted plants cannot cope with intense sunlight. They must be exposed in small increments. The use of shade cloth over the planting site aids in this endeavor. It can be gradually stripped away to allow the plant to adjust to the outdoor environment. Shade cloth also limits evaporation, moderates heat, and offers some wind protection.

STAKING AND TRELLISING Young trees and trees with poorly developed root systems may benefit from staking. The need is most acute in open areas prone to windy conditions and in loose, sandy soil where roots are unable to obtain a solid purchase. The material used to fasten the tree to the stake must not constrict the trunk. If guy wires are used, a tube or rubber hose should be used to prevent chafing. Once a tree is established—usually after the passage of a single growing season—the stake should be removed so that it does not interfere with normal development.

These pitayas are grown on a specialized support system, designed to encourage a spray of fruiting branches.

Four of the five vining plants profiled within this book—muscadine grape, kiwifruit, passionfruit, and pitaya—require a trellis or some other support system. Blackberries and other brambles may benefit from a single-wire trellis. A trellis maximizes the plant's exposure to sunlight, keeps fruit away from the ground, protects against attack by ground-dwelling pests, and makes the fruit easy to harvest. Although a chain link fence can serve as a temporary "trellis," such an arrangement may fail over the long term. Vines soon become intertwined with links. The result is often a cascade of tangled foliage and a ruined fence.

A trellis system is absolutely critical when raising grapes and most other varieties of vine-borne fruit.

FRUIT TREE PESTS

Many creatures that inhabit the garden are not pests. Lady beetles, assassin bugs, lacewings, Ichneumonid wasps, and praying mantises all play an important role in keeping other insect populations under control. Honeybees and various flies and beetles assist in pollination. When pesticides are applied needlessly or carelessly, they can wipe out these helpful insects. Pesticides should not be thought of as a "preventive measure." They should be applied only when insects begin to overrun a tree, when their presence can no longer be ignored.

With that said, Florida is home to an extraordinary number of pests—native and imported. Many are difficult or impossible to control. Every gardener can expect to lose fruit to one or more of the following animals.

CARIBBEAN FRUIT FLY The Caribbean fruit fly, *Anastrepha suspensa,* is an extremely destructive pest of many types of fruit crops. It is most prevalent in south Florida, but has been detected throughout the peninsula. It is abundant from March through September. The fly measures about a third of an inch in length. Its wings are marked with irregular dark-brown bands. The abdomen terminates in a spikelike ovipositor. White maggots hatch about two days after the female fly deposits its eggs under the skin of the fruit. They tunnel through the pulp and render the fruit unfit for human consumption. The list of species attacked by the Caribbean fruit fly is extensive. Chemical pesticides are ineffective. Several types of traps have been tested with mixed results. The only sure way to prevent infestation is to bag individual fruit.

These bags, made of breathable fabric, protect against the ravages of the Caribbean fruit fly.

GRAY SQUIRREL The eastern gray squirrel, *Sciurus carolinensis,* will eat seeds and nuts; however, its favorite food is fruit. It will sometimes knock down scores of fruit, take several bites out of each, and leave the ruined fruit to spoil. The squirrel seems to target the largest, sweetest, most desirable fruit in the grove. It will repeatedly risk its life to return to the tree that is the source of such delicacies. These destructive tendencies make the squirrel an extremely unwelcome visitor to the home orchard. Most commercial repellents have little effect. Traps can reduce populations but will not eliminate the problem. A permit is required to relocate the gray squirrel.

NEMATODES Nematodes are the most populous group of multi-celled animals on earth. Tens of thousands of these microscopic worms can exist within a cubic foot of soil. Young plants are particularly susceptible to nematode damage since their roots are relegated to the upper levels of soil where nematode populations are concentrated. Mulch can reduce the number of nematodes within the root zone.

BOAT-TAILED GRACKLE The boat-tailed grackle, *Quiscalus major,* is a common pest throughout much of Florida. This large bird is noisy and aggressive. Because of its black, iridescent coloration, it is often mistaken for a crow. The boat-tailed grackle consumes great quantities of lychee, longan, loquat, and other small-bodied fruit. Scare devices are somewhat effective, but will be ignored unless they are periodically moved or readjusted.

WEEVILS The Diaprepes root weevil, *Diaprepes abbreviatus,* is a serious pest of many fruit trees. It is about 1/2 inch in length, dark in color with bright orange stripes along its abdomen. The adult feeds on foliage, cutting rounded notches in leaf margins. The larva feasts on tree roots. An aggressive program may be required to control this pest. Oil sprays can reduce the deposit of eggs on foliage. Approved soil barrier pesticides may prevent the larva from penetrating the soil surface and damaging the roots.

The root weevil is exceedingly difficult to control.

SCALE INSECTS Scale insects feed on plant fluids. They attach themselves to the host plant, protected beneath tiny shells that superficially resemble miniature fish scales. Heavy infestations can reduce tree vigor or cause gradual decline. Scales can be controlled by sprays of oil and approved pesticides.

THRIPS Thrips, in particular the red-banded thrip, *Selenothrips rubrocinctus,* attack many fruit species in Florida. They are tiny and their presence is difficult to detect. They pierce the surface of foliage and fruit, causing stippled lesions and leaf deformation. Severe infestations can cause defoliation. Approved pesticides are somewhat effective in controlling this pest.

MITES Mites are tiny insects that cause widespread damage to fruit crops in Florida. The rust mite feeds predominantly on fruit surfaces, causing bronze-colored blemishes. The spider mite usually feeds on mature leaves. Broad-spectrum pesticides are often ineffective against mites and specially crafted miticides may be needed to suppress populations.

WHITEFLIES The adult whitefly resembles a tiny white moth and measures less than 1/6 inch in length. The larva feeds on plant juices. The whitefly accounts for significant agricultural losses. Infestations cause trees to lose vigor. In addition, this insect is responsible for transmitting several plant viruses. This pest is difficult to control with broad-spectrum pesticides. Insecticidal oils or soaps provide some relief. Multiple applications may be necessary.

EASTERN COTTONTAIL RABBIT The eastern cottontail rabbit, *Sylvilagus floridanus,* is found throughout Florida. While it will eat fruit that dangles within its reach, it poses a far more serious threat. The rabbit is capable of girdling a fruit tree by gnawing at the bark. It is especially attracted to members of the Roseaceae family, such as the apple. The best way to prevent rabbit damage is to erect a fence or to wrap the lower trunk with protective material.

ROOF RAT The roof rat, *Rattus rattus,* attacks banana, citrus, lychee, papaya, pineapple, and other fruit crops, often by hollowing out the fruit interior. The roof rat can be killed with a trigger trap. However, such traps should only be set during the nighttime to avoid injury to birds and other species. Several poison baits are also approved for combating this pest.

APHID The aphid is an extremely common pest of many fruit trees and ornamentals. This pear-shaped insect is tiny, measuring less than 1/8 inch in length. It congregates on the underside of developing leaves. It can form dense colonies, sapping the energy of the plant and causing the distortion of new growth. Populations tend to peak in the spring. The aphid is relatively easy to control. Insecticidal oils and soaps are effective and usually have relatively low toxicity.

FRUIT TREE DISEASES

Plant diseases manifest themselves through many symptoms, including leaf blight, shoot blight, panicle blight, rots, cankers, leaf spots, wilts, limb dieback, and general dieback. They are often difficult to diagnose and can easily be confused with nutritional deficiencies or pest-related damage. The discussion that follows is limited to a few common diseases with well-recognized symptoms.

ANTHRACNOSE Anthracnose is an important fungal disease of mango, avocado, and other fruit crops. It often manifests itself as dark, spreading lesions on the skin of fruit. These lesions coalesce and eventually spread decay to the interior of the fruit. Spores may be transferred by rainwater dripping from dead wood or other infected parts of the plant. The disease is most prevalent during periods of warm, rainy weather. Anthracnose can be controlled through the use of fungicidal sprays.

The black spots on this mango are symptoms of anthracnose. At this early stage of infection, the fruit is still fit to eat.

PHYTOPHTHORA ROOT ROT Phytophthora root rot is a common fungal disease of fruit trees in Florida and elsewhere. The fungus lies dormant in the soil until activated by wet weather. It then infects root tissue. If left untreated, root rot will generally kill the host plant. Symptoms include a reddish discoloration of the inner bark at the base of the trunk, along with widespread necrosis of the outer roots. The disease can be avoided by planting susceptible species in high, well-drained sites. If the disease is recognized at an early stage, approved fungicides, applied as a soil drench, may save the tree.

FIREBLIGHT Fireblight is a serious disease that affects members of the Rosaceae family. It is caused by the bacterial pathogen *Erwinia amylovora*. The disease is spread by insects and is triggered by persistent wet conditions. The classic symptom is that an entire limb will experience sudden dieback, giving the appearance of having been scorched by fire. In addition, oozing cankers form in the bark of infected trees.

Fireblight is difficult to control. Infected limbs should be severed well below any portion exhibiting signs of infection. Pruning shears or other tools must be disinfected after each cut. Severed branches should be burned or bagged and removed from the orchard.

CITRUS CANKER Citrus canker is a disease caused by the bacterial pathogen *Xanthomonas axonopodis*. Infection results in raised brown lesions on fruit, leaves, and stems, followed by premature fruit drop and general decline. The disease is highly contagious and can be spread by garden shears, landscape equipment, wind-driven rain, animals, and people. No treatment exists. The only known control is the destruction of an infected tree and other trees in the vicinity. Eradication efforts have been only marginally effective and have resulted in numerous quarantines and the removal of millions of trees.

CITRUS GREENING Citrus greening, caused by the bacteria *Liberibacter asiaticus,* is a new disease in Florida. It has caused extensive damage in other regions of the world. Insects transmit the disease from one plant to another. Symptoms include blotchy yellowing of leaves, lopsided fruit that drop prematurely before coloring, followed by gradual decline and the death of the tree. No effective controls exist. Citrus greening represents a very serious threat to commercial and dooryard citrus crops in Florida.

VERTICILLIUM WILT Verticillium wilt, a fungal disease, attacks a wide range of fruiting plants. The fungus can lay dormant in the soil for decades. When conditions are right, it invades the roots of the plant and spreads upward, destroying the plant's vascular tissue. Infection is hard to detect until substantial damage has occurred. The most recognized symptom is dieback along one side of a tree. Verticillium wilt is common on land formerly used for the production of vegetables. Where the disease is present, the grower should plant resistant species and should take extra care to ensure adequate drainage.

CERCOSPORA LEAF SPOT Several fungi within the genus *Cercospora* inflict damage on the foliage of fruit trees. Spores from infected plants are dispersed by wind and rain. Warm, wet conditions accelerate the spread of the disease. Symptoms include circular spots with gray centers and dark brown to reddish-brown margins. Infected leaves turn bronze to light brown, then grayish. The disease first affects older leaves. Inner branches may be denuded, leaving only clusters of leaves near the branch tips. The disease can be controlled through the use of fungicide sprays.

POWDERY MILDEW Powdery mildew can be caused by several fungi and is prevalent on many crops. The disease manifests itself in the form of white lesions appearing on foliage, whitish-gray discoloration of twigs and emerging growth, aborted blossoms, and blemished fruit. It may lead to loss of tree vigor and reduction in yield. Several fungicides are effective at treating the disease.

ALGA SPOT Alga spot, sometimes referred to as red rust, is a relatively minor disease, caused by the algae *Cephaleuros virescens*. Symptoms appear as pale green circular spots on leaves, which eventually turn reddish-brown. Alga spot is most prevalent during periods of wet weather and may be transferred from one tree to another by wind-driven rain. The grower can prune out infected branches to prevent further spread.

SOOTY MOLD Sooty mold is not really a disease, as it does not directly attack plant tissue. However, its effects are similar. The fungi that cause the disorder are associated with the liquid exudates of various small insects, including aphids, mealybugs, scales, and whiteflies. Sooty mold shows up as a black, velvety coating on the surface of leaves. Once the underlying insect problem has been solved, it will dry up and flake off. Sooty mold is unsightly, but is rarely a serious threat to the health of the tree.

COLD PROTECTION

Protecting fruit trees from cold winter temperatures is an important consideration for growers throughout mainland Florida. From December through March, gardeners should carefully monitor weather predictions. When cold temperatures threaten, initial efforts should be directed toward protecting young plants, valuable specimens, and species with the least cold tolerance.

Long before cold weather arrives, the fruit gardener should take this threat into account, both in the selection of plants and in the location of plants within the garden. Growers should take advantage of local microclimates. Significant temperature differences can occur even within the confines of a small and relatively level suburban lot. The grower should attempt to visualize the movement of cold air across the property. Fences, hedges, buildings, stands of trees, and other raised objects can create a barrier to the passage of cold air. Cold air drains away from elevated areas and collects in depressions and low areas. Cold-sensitive plants should be planted in high or protected locations. Cold-hardy plants should be placed in open or low locations.

When properly deployed, under canopy sprinklers or overhead sprinklers can protect cold-sensitive plants. Water protects plants in several ways. Every gram of water that turns to ice releases about 80 calories of heat. In addition, the temperature of water drawn from the ground by irrigation systems may be more than 30°F warmer than the air temperature. Finally, ice that forms over plant tissue forms a protective barrier against cold temperatures. The rate of application must be increased to compensate for lower temperatures and greater wind speed.

A critical consideration when faced with a severe freeze is to save the trunk of the tree. Twig and foliage damage is usually temporary. Damage to the trunk or graft union may lead to the death of the tree. By wrapping the trunk with heavy insulation down to ground level, growers can help ensure that new growth will sprout when warmer temperatures return. Temporarily banking mulch against the trunk may also provide some measure of protection.

Some growers use smudge pots or other heat-producing devices to protect against cold-weather injury. However, the effectiveness of such systems is greatly reduced by windy conditions. Consequently, many growers have abandoned smudge pots in favor of overhead sprinklers. Smaller trees can be enclosed in frames wrapped with plastic or cloth. Lights placed within such enclosures can raise the temperature by a few degrees, adding an additional layer of protection.

Many factors play a role in determining the extent of freeze-related injury. These include the lowest temperature reached, the duration of the freeze, the overall health of the tree, the amount of wind, the exposure of the planting site, and the effectiveness of protective measures. Once the cold snap has passed, fruit gardeners should hesitate before trimming away damaged limbs or ripping out "dead" trees. Many cold-damaged trees begin putting out green shoots after two or three months of apparent lifelessness.

WIND PROTECTION

Florida is periodically raked by tropical storms and hurricanes. The state's lengthy coastline and exposed position make it especially vulnerable. These storms cause immense damage to groves, orchards, and home gardens.

The selection of wind-resistant plants can limit hurricane damage. Such plants tend to have a low center of gravity, flexible wood, a compact profile, and a deep penetrating root system. Trees with good resistance to hurricane-force winds include akee, atemoya, carambola, Darling plum, feijoa, grumichama, Kei apple, mamey sapote, mulberry, rumberry, sapodilla, tamarind, and white sapote. As a rule, established trees fare better than recently planted trees.

Although stripped of its fruit and leaves by hundred-mile-per-hour winds, this carambola suffered no lasting injury.

The use of sturdy trees as windbreaks along the property border can reduce hurricane damage. Proper pruning techniques can also play a vital role. Pruning to encourage the growth of strong, wide-angled scaffold limbs can reduce breakage. The time to prune is well before the approach of a storm. As a storm nears, it is likely that the homeowner will be preoccupied with other tasks. Topping trees is sometimes considered a drastic measure. However, many species respond well if the upper quarter of the canopy is removed on an annual or biannual basis. This technique can also create thicker, bushier growth. The wind can gain great leverage over tall, sail-like trees, so size control is critical.

Leaves play a key role when it comes to hurricane damage. They act like thousands of tiny sails pulling at the tree and its roots. If it appears that a direct hit from a storm is inevitable, the homeowner may be able to save a small or valuable specimen by removing the leaves with a hedge trimmer or by hand. Presuming the tree is in good health, it will quickly put out new growth once the storm has passed. Obviously, human safety should never be compromised in an effort to save a fruit tree.

This jaboticaba was knocked over by hurricanes three times in two years. It was righted, staked, heavily watered, and never missed a production cycle.

Coconuts should be removed from a coconut palm as soon as possible when a storm threatens. They can be turned into the rough equivalent of cannonballs by hurricane-force winds. Other fruit that is nearing maturity should be picked in advance of the approach of a hurricane or tropical storm as it will otherwise be slung off and lost.

AFTER THE STORM Caring for trees soon after the storm passes is crucial. Many toppled trees can be righted. However, if exposed roots are permitted to dry out, any hope of saving a tree may be lost. If the root system has been compromised, trees should be braced or staked. When roots have been badly damaged, it may be necessary to trim away portions of the canopy to balance the tree and reduce demands on the root system. The area around any exposed roots should be thoroughly backfilled. The tree should be regularly irrigated for a period of several months.

Limb damage is a less immediate concern, but must also be addressed within a reasonable time. Shattered limbs should be removed. Wounds caused by severed limbs should be coated with asphalt or other sealant. The grower may be able to save split branches by inserting screws to bridge the gap. These stabilize the wound, giving the branch an opportunity to heal. With some effort, most trees can be salvaged from damage inflicted by hurricane winds.

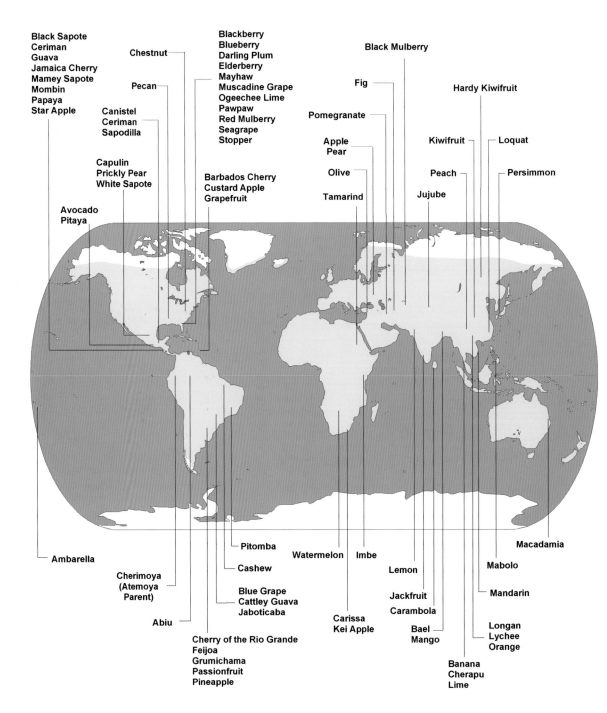

Black Sapote
Ceriman
Guava
Jamaica Cherry
Mamey Sapote
Mombin
Papaya
Star Apple

Chestnut

Pecan

Canistel
Ceriman
Sapodilla

Capulin
Prickly Pear
White Sapote

Avocado
Pitaya

Blackberry
Blueberry
Darling Plum
Elderberry
Mayhaw
Muscadine Grape
Ogeechee Lime
Pawpaw
Red Mulberry
Seagrape
Stopper

Barbados Cherry
Custard Apple
Grapefruit

Black Mulberry

Fig

Pomegranate

Apple
Pear

Olive

Tamarind

Hardy Kiwifruit

Kiwifruit

Loquat

Peach

Persimmon

Jujube

Ambarella

Cherimoya
(Atemoya
Parent)

Abiu

Cherry of the Rio Grande
Feijoa
Grumichama
Passionfruit
Pineapple

Pitomba

Cashew

Blue Grape
Cattley Guava
Jaboticaba

Watermelon

Imbe

Lemon

Jackfruit
Carambola

Carissa
Kei Apple

Bael
Mango

Macadamia

Mabolo

Mandarin

Longan
Lychee
Orange

Banana
Cherapu
Lime

This map shows the approximate geographic origin of the fruiting plants profiled within this book.

18

II

Fruit Profiles

Abiu

SCIENTIFIC NAME: *Pouteria caimito*
FAMILY: Sapotaceae
OTHER COMMON NAME: Caimito

Fruiting Calendar

JAN	FEB	MAR	APR	MAY	JUN	JUL	AUG	SEP	OCT	NOV	DEC

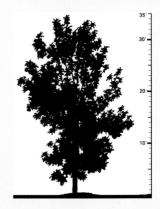

Characteristics

Overall Rating	★★★
Ease of Care	★★★
Taste/Quality	★★★★
Productivity	★★★★
Landscape Value	★★★
Wind Tolerance	★★★★
Salt Tolerance	★★
Drought Tolerance	★★
Flood Tolerance	★★★★
Cold Tolerance	★

Until recently, many considered the abiu a minor tropical fruit. Its prospects have brightened with the discovery of superior varieties and with its emergence as a commercial crop in Australia. The fruit is attractive and is of excellent quality. Dooryard growers are beginning to recognize the merits of this delicious, low-acid fruit. The tree is relatively pest free and is capable of bearing heavy crops. Unfortunately, the abiu is rare in Florida, where its intolerance of winter cold may prove a limiting factor.

Known Hazards

None

GEOGRAPHIC DISTRIBUTION The abiu originated in the western Amazon Basin. The species was introduced into Florida in 1914, but failed to arouse much interest until recent times.

TREE DESCRIPTION The abiu is a medium, evergreen tree, resembling its cousin, the canistel, in overall shape and habit of growth. The ultimate size of the tree in Florida has not been established, although mature specimens typically measure 30 feet or more. Like many species within the Sapotaceae family, the abiu exudes sticky latex when twigs are broken or when the bark is cut. The leaves are alternate, glossy, elliptic, and measure between 4 and 9 inches in length. Inconspicuous bell-shaped flowers are born in the leaf axils. The abiu is a moderately fast-growing tree. A grafted specimen will typically begin to bear after three or four years.

FRUIT CHARACTERISTICS The fruit is light yellow when ripe, with smooth, glossy skin. The abiu is variable in size and shape. It ranges from spindle-shaped to oval and sometimes has a distinct nipple at the apex. While typically the size of a large hen's egg, the fruit may measure from 2 to 4 1/2 inches in length. It may weigh from a few ounces to 1 1/4 pounds. In better cultivars the pulp is sweet, soft, and mild, hinting of caramel and cream. One to 4 brown oblong seeds surround the core in a radial pattern.

CULTIVARS Those who seek to grow the abiu should attempt to obtain a grafted specimen. Unfortunately, few nurseries in Florida carry this species and selection is limited. The best variety thus far introduced into the United States is probably the 'Gray' cultivar. Rare-fruit pioneer William Whitman brought this tree from Australia in 1991. It produces a 4-inch fruit with superior flavor. Other cultivars sometimes available are 'Lu,' 'Amazon,' and 'Z4.' In Australia, a host of cultivars exist, including 'Cape Oasis,' 'Inca Gold,' 'T25,' 'T31,' 'Z1,' 'Z2,' 'Z3,' and 'Z4.'

RELATIVES The Sapotaceae or Sapote family, within the order Ebenales and subclass Dilleniidae, includes about 50 genera and 1,000 species. It is extremely rich in fruit-bearing members. Other members of the *Pouteria* genus include the canistel, *Pouteria campechiana,* and mamey sapote, *Pouteria sapota,* both described in this book. The abiu is more distantly related to the sapodilla, *Manilkara zapota,* and star apple, *Chrysophyllum cainito,* also described in subsequent sections.

CLIMATE The abiu prefers a high, even temperature, high humidity, and high rainfall evenly distributed throughout the year. In keeping with its tropical origin, the abiu is cold sensitive. A temperature drop to 30° F will cause defoliation. A lower temperature may kill the tree outright.

CULTIVATION The abiu is a low-maintenance tree. It performs satisfactorily on most types of soil, including clay and sand, but it seems to do best on soils with a high water-holding capacity. It requires irrigation during fruit development and during periods of drought. Little information exists on the tree's nutritional requirements, although growers can safely assume that light applications of a balanced fertilizer will benefit growth and enhance productivity. To date, no significant pests or diseases have affected the abiu in Florida. Superior cultivars are grafted or air layered.

HARVEST AND USE In Florida, fruit mature from August through October. A mature specimen may produce as much as 400 pounds of fruit. The fruit should be harvested when it is still firm, but after most of the skin has turned yellow. If the fruit is picked prematurely, latex will interfere with eating quality. It is considered ready to eat when the skin is light yellow. By the time the skin turns golden yellow, the fruit is past its prime. The abiu is best eaten chilled. The pulp can simply be spooned from the rind. Once cut, the fruit discolors with exposure to air.

Until it reaches maturity, the fruit of the abiu is green, hard, and full of acrid latex.

Ambarella

SCIENTIFIC NAME: *Spondias dulcis*
FAMILY: Anacardiaceae
OTHER COMMON NAMES: Otaheitte Apple, June Plum,
  Ciruela (Spanish)

Fruiting Calendar

JAN	FEB	MAR	APR	MAY	JUN	JUL	AUG	SEP	OCT	NOV	DEC

Characteristics

Overall Rating	★★
Ease of Care	★★★★
Taste/Quality	★★
Productivity	★★★★
Landscape Value	★★★
Wind Tolerance	★★★
Salt Tolerance	★★
Drought Tolerance	★★★
Flood Tolerance	★★★
Cold Tolerance	★

The ambarella grows well in coastal areas of south Florida and produces a fruit of fair quality. The flavor is pleasant and the juice refreshing, but the spiny core and tough fibers detract from eating quality. In texture and taste, the fruit has been likened to an under-ripe pineapple. Nevertheless, the ambarella has several advantages. It produces over the fall and winter when few other species are in season. It bears ample quantities of fruit over several months. The tree is attractive and requires little care.

Known Hazards

The fruit may present a choking hazard if the core and fibers are not properly removed. Contact with foliage and sap may cause dermatitis in some individuals.

GEOGRAPHIC DISTRIBUTION The ambarella is native to Polynesia. It is grown as a commercial crop throughout Southeast Asia. The ambarella was introduced into Florida in the early 1900s. Since that time, it has been planted primarily as a curiosity and is occasionally included in rare fruit collections. It has not attained commercial import in Florida.

TREE DESCRIPTION The ambarella is a medium, deciduous tree. It reaches a height of about 40 feet in south Florida. The leaves are ornamental and compound, made up of between 9 and 25 small leaflets with toothed margins. These average slightly over an inch in length. The trunk is gray and smooth. The wood is relatively soft; the branches are stiff and brittle. The ambarella is fast growing and seedlings may fruit in as little as 3 years. The tree has good landscape value and serves as a small shade tree—except during its winter dormancy period. Small, white flowers are borne on terminal panicles. Each panicle supports a combination of male, female, and perfect flowers.

FRUIT CHARACTERISTICS The fruit of the ambarella is an oval drupe measuring between 2 and 2 3/4 inches in length. Long stalks bear pendant clusters of up to 30 fruit. As the fruit ripens, the skin turns from green to yellow to golden-orange. The skin is tough and inedible. The fruit is best when fully colored, but still somewhat crunchy. At this stage, it has a pineapple-mango flavor. The flesh is golden in color, very juicy, vaguely sweet, but with a hint of tart acidity. When soft and overripe, the fruit takes on an unappealing musky character. At all stages, the flesh is laced with a network of several tough fibers or cords that run through the fruit longitudinally. These surround a spiny, inedible core, which holds from 1 to 5 seeds.

CULTIVARS While no named cultivars are available, a dwarf tree is sold by several Florida nurseries. It only attains a height of 10 or 12 feet and will grow in a container for many years.

RELATIVES The Anacardiaceae or Spondia family, within the order Sapindales and the subclass Rosidae, consists of about 70 genera and 600 species. Important fruit-bearing species include the mango, *Mangifera indica,* and cashew, *Anacardium occidentale.* Other noteworthy species are discussed within the subsection pertaining to relatives of the mango.

CLIMATE The ambarella is suited to a humid tropical climate. It is slightly more sensitive to cold than the mango, and prefers a frost-free location. Temperatures of less than 29° F may cause significant damage.

CULTIVATION The ambarella is a low-maintenance tree. It is not particular as to soil type but, like most tropical fruit species, cannot stand wet or poorly drained soils. The tree is drought tolerant and, like most species within the *Spondias* genus, prefers a pronounced dry season. A young tree may benefit from periodic irrigation until it is established. Thereafter, no irrigation is necessary. Apart from removing dead wood or the occasional crossing branch, the tree requires no pruning. If branches are cut back severely they will not fruit the next season. The tree benefits from periodic applications of a balanced fertilizer. It grows well in protected locations and does not seem to mind partial shade. It is relatively free of disease and insect pests. Seeds germinate readily. The tree can also be reproduced through air layering and hardwood cuttings.

HARVEST AND USE In Florida, the ambarella fruits over a lengthy period from September through March. The fruit may be harvested when fully sized but still green, or after it has turned yellow. When still green, the fruit can be used in salads, curries, or pickles. As a result of its complex interior structure, the fruit does not lend itself to easy preparation. It can be eaten raw, sliced, juiced, or used in sauces.

Blooms are followed by clusters of young fruit.

Apple

SCIENTIFIC NAME: *Malus pumilla* or *Malus domestica*
FAMILY: Rosaceae
OTHER COMMON NAME: Manzana (Spanish)

Fruiting Calendar

JAN	FEB	MAR	APR	MAY	JUN	JUL	AUG	SEP	OCT	NOV	DEC

Characteristics

Overall Rating	★★
Ease of Care	★★★
Taste/Quality	★★★
Productivity	★★★
Landscape Value	★★★
Wind Tolerance	★★★
Salt Tolerance	★★
Drought Tolerance	★★★
Flood Tolerance	★★★
Cold Tolerance	★★★★★

The apple is poorly adapted to Florida growing conditions. The best northern cultivars do not receive sufficient chilling hours to produce fruit. However, to meet the needs of those determined to raise temperate fruit in warm climates, several low-chill cultivars have been developed. These may not measure up to the best northern varieties, but they bear fruit of good quality and are reasonably productive in north Florida.

Known Hazards

The seed contains toxins.

GEOGRAPHICAL DISTRIBUTION The domesticated apple originated in Asia Minor, probably in Armenia, Georgia, or eastern Turkey. It made its way to the Mediterranean and Europe prior to 100 B.C. Early colonists introduced the apple into North America. Several *Malus* species native to North America produce edible fruit; however, none are considered to have commercial potential.

TREE DESCRIPTION The apple is a small to medium deciduous tree, growing to a height of about 20 to 25 feet. The dull-green leaves are elliptic with serrate margins. Flowers, which typically form in clusters of from 4 to 6, are light pink, becoming white when fully open. Most cultivars require cross-pollination from other cultivars to set fruit. The crab apple can serve as pollinator.

FRUIT DESCRIPTION The apple is a pome. The flesh develops from the hypanthium or floral cup, not from the ovary of the flower. The actual "fruit" is the inedible core. The core contains 5 cavities, each of which contains 1 or 2 small, black seeds.

CULTIVARS Of the more than 2,000 cultivars of apple that have been selected, only a handful will fruit adequately in Florida. 'Anna' is an Israeli selection that requires 250 to 300 chilling hours. It regularly fruits as far south as Orlando. 'Golden Dorsett,' a 'Golden Delicious' seedling that originated in the Bahamas, probably has the lowest chilling requirements of any selection. It fruits sporadically even in southern parts of the peninsula. Other low-chill varieties include 'Tropic Sweet,' a patented variety developed by the University of Florida, and Israeli selections 'Ein Shemer,' 'Michal,' 'Maayan,' and 'Shlomit.'

RELATIVES The large and diverse Rosaceae or Rose family, within the order Rosales and the subclass Rosidae, is thought to include as many as 3,000 species. Many important temperate fruit species, along with a scattering of tropical species, belong to this family. The apple belongs to the Pomoideae subfamily, which includes the loquat, *Eriobotrya japonica;* mayhaw, *Crataegus* spp.; and pear, *Pyrus* spp., each profiled within this book. Other Rosaceae species presented within these pages include the blackberry, *Rubus* spp.; capulin, *Prunus salicifolia;* chickasaw plum, *Prunus angustifolia;* peach, *Prunus persica;* and strawberry, *Fragaria ananassa.*

CLIMATE The tree is not affected by winter cold in Florida. However, flowers may be injured by late frosts. As previously noted, the tree must receive significant exposure to temperatures below 45° F to stimulate flowering and fruit set. Intense heat and high humidity are not conducive to fruit production.

CULTIVATION The apple is a moderately demanding tree. It prefers well-drained loam with a pH between 6 and 7, but can adapt to a wide range of soil types. It prefers full sun, but can endure some shade during the early morning or late afternoon. Competing vegetation should be eliminated from around the base of a young tree. The apple should be pruned to a vase structure or modified central leader system. Fruit should be thinned to leave one per spur. Rabbits will gnaw at the bark and can readily girdle a tree. Trunk wraps or other barriers may be required. As with several other members of the Roseceae family, the species is susceptible to fireblight. Botryosphaeria canker is a serious fungal disease affecting production in Florida. Treatment is similar to that for fireblight, in that infected branches must be pruned out and disposed of in a hygienic fashion. The apple is propagated through various grafting techniques, with most resulting in a high rate of success.

HARVEST AND USE In north Florida, low-chill cultivars ripen from May through early July. Maturity is indicated by a combination of factors: a slight softening, a few fallen fruit and a change in background color. The apple is one of only a handful of fruit species that can endure extended storage without an appreciable loss in quality.

A 'Golden Dorsett' apple matures in Homestead, proof that temperate species will occasionally bear fruit under near-tropical conditions.

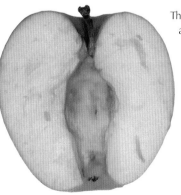

The true fruit of the apple is the core. The surrounding material—usually thought of as the fruit—is the swollen floral cup.

Apple

Atemoya

SCIENTIFIC NAME: *Annona cherimola* x *Annona squamosa*

FAMILY: Annonaceae

Fruiting Calendar

JAN	FEB	MAR	APR	MAY	JUN	JUL	AUG	SEP	OCT	NOV	DEC

Characteristics

Overall Rating	★★★★
Ease of Care	★★★
Taste/Quality	★★★★★
Productivity	★★
Landscape Value	★★★
Wind Tolerance	★★★
Salt Tolerance	★
Drought Tolerance	★★★
Flood Tolerance	★★★
Cold Tolerance	★★

A premium fruit with few equals, the atemoya is on the verge of becoming a popular dooryard fruit in south Florida. It is a member of the venerable Annona genus. Scarceness, superb flavor, and overall quality make this species a valuable addition to any home garden. The tree has some cold tolerance, although growth is generally limited to the southern half of the peninsula. The atemoya is attacked by several insect pests and tends to be a shy bearer absent pollination by hand.

Known Hazards
The seed contains potent toxins.

GEOGRAPHIC DISTRIBUTION The atemoya is a cross between the cherimoya and the sugar apple. The cherimoya is often regaled as the most delectable fruit of the western hemisphere. A native of the slopes and high valleys of the Andes Mountains, it can withstand temperatures down to about 25° F. However, it does not tolerate high summertime temperatures. The sugar apple requires a near-tropical climate. In many respects, the atemoya merges the best qualities of both species. The first atemoya was bred in Miami in 1908. The species is grown as a minor commercial crop in south Florida.

TREE DESCRIPTION The atemoya is a small tree reaching about 20 feet in height. It is briefly deciduous, dropping its leaves for 2 months in late winter. The branches are sparse. The canopy is sprawling and is roughly globose in form. The tree has smooth gray-brown bark. Leaves are alternate, fleshy, ovoid, and measure between 3 and 7 inches in length. The tree grows relatively quickly and may begin to fruit within 3 years after it is planted. The tree has fair landscape value.

FLOWERING AND POLLINATION The flower measures about 1 1/2 inches in length. It has 3 fleshy petals, green on the outside and light green on the inside. Flowers are born singly on thin stems. Flowering occurs throughout the spring. The tree may also have a minor late-summer bloom.

The female flower as it appears shortly before it spreads its petals and becomes a male.

The flower is protogyrus. When the flower first emerges it is female. However, toward the end of its existence, the petals spread, revealing a ring of stamens coated with pollen. At this stage the female parts are no longer receptive to pollen. Wet conditions during flowering tend to enhance fruit set.

Most cultivars require hand pollination to set adequate quantities of fruit. A narrow, artist's brush and a film canister are the only tools required. The best time to perform this operation is between 4:00 P.M. and dusk. The key to success lies in finding male flowers in the process of shedding pollen. The pollen should be carefully brushed from the male flower into the container. Pollen loses viability quickly, but may be stored in the refrigerator for short periods.

Female flowers are more numerous. Because of the narrow opening between the petals, pollinating the female flower can be tricky. It may be necessary to break away a petal to gain access. The brush should be used to gently rub pollen across the surface of the ovary. This procedure should be repeated every few days in an attempt to pollinate as many female flowers as possible.

FRUIT CHARACTERISTICS The fruit is aggregate, ovate, and conelike toward the apex. The skin is made up of multi-sided areoles, which form fleshy protuberances. The surface presents a hatched, knobby pattern, causing the fruit to superficially resemble a fused artichoke. The fruit may weigh anywhere from 6 ounces to over a pound, and may measure from 3 to 6 inches in diameter. It matures 4 to 6 months after bloom. Skin color varies from blue-green, to pea green, to grayish-pink. The pulp is usually white, but may be tinged with pink. Poorly defined pulp segments radiate outward from a narrow, tapering central core. The atemoya combines an array of flavors: vanilla, wintergreen, and pineapple. The intense sweetness is offset by a hint of fruitiness and acidity. The texture is fine and crisp, yet moist and melting. The fruit may contain from 10 to 50 shiny, dark brown seeds.

CULTIVARS A number of cultivars have been selected in Florida, Israel, South Africa, Australia, and elsewhere. Great potential exists for further improvement.

'Gefner' - This Israeli selection produces fruit of excellent quality. It is the foremost commercial variety in Florida. It produces heavy crops without hand pollination. The flavor is superb. The only

The 'Gefner' cultivar performs well under Florida conditions.

drawback to 'Gefner' is that the fruit tends to be somewhat small and seedy.

'4826' - This Florida cultivar is a cross between the red sugar apple and the cherimoya. The fruit is pinkish-gray in color. The quality is very good. The tree is not particularly vigorous and is prone to limb dieback. Leaves are small. Nevertheless, '4826' can set respectable quantities of fruit without hand pollination.

'Bradley' - This cultivar rarely sets fruit without hand pollination. While the fruit is of very good quality, it tends to split on the tree. 'Bradley' was developed in south Florida.

'Priestly' - This cultivar bears large fruit of good quality. 'Priestly' will set a few fruit without hand pollination, but naturally pollinated fruit tend to be misshapen. Production is light. The tree is vigorous and has very large leaves.

'African Pride' - This cultivar is the primary commercial variety in South Africa and Australia. Along with 'Gefner' and '4826,' it is one of only a handful of cultivars that set reasonable quantities of fruit without hand pollination. Fruit quality is variable.

'Pink's Mammoth' - 'Pink's Mammoth' is considered the premium-quality atemoya in Australia. The fruit is very large. It has a low seed count and exceptional flavor. The tree is large and is a shy bearer. This cultivar does not appear to be widely available in Florida.

'Page' - 'Page' was one of the first cultivars developed in south Florida. Like 'Bradley,' the fruit is of small to medium size, and of good quality.

Other cultivars are being evaluated. These include Israeli selections: 'Kabri,' 'Jennifer,' 'Malamud,' 'Bernitski,' and 'Malalai.' Australia has an extensive breeding and selection program. 'Island Gem,' 'Maroochy Gold,' and 'Maroochy Red' have exhibited desirable characteristics. Other Australian clones include 'Martin,' 'Palethorpe's Pride,' 'Ruby Queen,' 'Island Beauty,' 'Lindstrom,' 'Nielsen,' and 'Paxton.'

RELATIVES The Annonaceae or Annona family, within the order Magnoliales and the subclass Magnoliidae, encompasses about 75 genera and 2,200 species, many of which are native to tropical America. This book devotes separate sections to the custard apple, *Annona reticulata;* the pawpaw, *Asimina triloba;* and the sugar apple, *Annona squamosa.*

Many regard the cherimoya, *Annona cherimola,* as the premium fruit within the Annona family. While it does not prosper in Florida's heat and humidity, it will grow in south Florida and will occasionally bear fruit. The fruit is large, conical or heart-shaped, and ranges in flavor from delicious to exceptional. Some cultivars may be more suitable to Florida growing conditions than those thus far attempted. California cultivars include 'Bays,' 'Big Sister,' 'Chaffey,' 'Deliciosa,' 'Honeyhart,' 'Knight,' 'Libby,' 'Lisa,' 'Ott,' 'Nata,' 'Santa Rosa,' 'Selma' 'Villa Pink,' 'Whaley,' and 'White.'

The ilana, *Annona diversifolia,* is highly regarded in its Central American homeland and is grown on a limited basis in South Florida. The fruit resembles that of the cherimoya. Quality can be exceptional. The thick rind reportedly limits attacks by the chalcid wasp—the bane of Annona growers in South Florida. Unfortunately, the tree is a light bearer. Recommended cultivars include 'Efrain' 'Fairchild,' 'Guillermo,' 'Imery,' 'Gramajo,' and 'Rosendo Pérez.'

The ilama, an obvious relative of the atemoya, is also a premium fruit well suited to growth in south Florida.

The biriba, *Rollinia deliciosa,* is a fruit of the Amazon basin. It will grow and fruit in south Florida but is cold sensitive. The fruit is heart-shaped, bright yellow, with pointed projections extending from each quasi-hexagonal segment. The pulp is mucilaginous, translucent and dotted with hundreds of small, dark seeds. The flavor is sweet with a hint of lemon.

The biriba resembles the atemoya, but is different in flavor, texture, and internal arrangement. The fruit turns yellow as it matures.

The soursop or guanabana, *Annona muricata,* is indigenous to northern South America, Central America, and the West Indies. It can be grown in coastal areas of extreme south Florida. The large fruit has a distinctive flavor and is greatly admired throughout the world's tropics. The pulp is often canned or processed into juice or other products.

The soursop is the only Annona that lends itself to processing and canning. The pulp is a popular ingredient in milkshakes and ice cream.

The pond apple, *Annona glabra,* is native to south Florida and the West Indies. It prefers wetland habitats. While the fruit is edible, the flesh is of poor quality. The tree is sometimes used as a rootstock for other Annonas. In theory, this arrangement should permit production in flood-prone areas. However, in practice, the graft union is prone to galling and other defects.

While the flower and fruit of the native pond apple resemble those of the soursop, the fruit is only of marginal edibility.

CLIMATE Winter cold is a limiting factor. The tree will survive brief temperature drops to 26° or 27° F. However, it should probably not be attempted north of Merritt Island, on the east coast, or Sarasota, on the west coast.

CULTIVATION The atemoya requires moderate care. It will tolerate a wide range of soils, and appears to prosper in the sand found across much of the Florida peninsula. It prefers a pH of 6.0 to 6.7. While the atemoya can survive brief flooding, it is susceptible to root rot in poorly drained locations. The atemoya benefits from full sun. At the same time the tree should be protected from strong wind. Its vulnerability to wind damage increases

when it is heavily laden with fruit. Growers often train the tree to send up a single straight shoot until it reaches about 40 inches in height. At that point, the terminal shoot is pinched to encourage lateral branching.

The atemoya is not particularly demanding when it comes to water, but does require periodic irrigation during periods of flowering, vegetative growth, and fruiting. A young tree should be watered twice a week for 3 or 4 months until it is established. A mature tree should be watered once a week during periods of active growth. Irrigation should be reduced or discontinued during the late fall and early winter. A mature tree should receive about a pound of 8–3–8 fertilizer 3 times annually.

PESTS AND DISEASES The chalcid wasp is an extremely vexatious pest. It was introduced into Florida in the 1920s. The wasp lays eggs in immature fruit. The larvae pupate within the seed. About 3 months later, the adult tunnels through the pulp, emerges through the rind, and flies off to ruin other fruit. Apart from bagging individual fruit at an early stage of development, few effective controls exist. Other insect pests include ambrosia beetles, mealy bugs, and thrips. The atemoya is susceptible to anthracnose, which primarily affects the ripening fruit.

PROPAGATION Seeds typically germinate in just over a month. Most grafting techniques are successful. Scions are taken from the previous year's growth.

HARVEST AND USE In Florida, fruit ripen during the late summer and fall. The atemoya is moderately productive. A mature 'Gefner' tree can bear upwards of 100 fruits. The fruit may undergo a subtle color change or may exhibit some rubbery give when it is ready to be harvested. It should be clipped from the tree, leaving a short piece of stem attached to the base.

The fruit should not be allowed to soften on the tree. Overripe fruit will split or develop black spots associated with anthracnose. Minor discoloration or blackening of the skin does not affect flesh quality. The fruit is eaten when it becomes slightly soft to the touch. The atemoya is best used as a fresh fruit, served halved and spooned from the skin. Prolonged refrigeration will result in chilling injury.

Avocado

SCIENTIFIC NAME: *Persea americana*
FAMILY: Lauraceae
OTHER COMMON NAME: Aguacate (Spanish)

Fruiting Calendar

JAN	FEB	MAR	APR	MAY	JUN	JUL	AUG	SEP	OCT	NOV	DEC

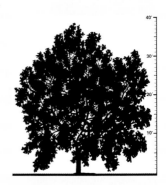

Characteristics

Overall Rating	★★★★★
Ease of Care	★★★
Taste/Quality	★★★
Productivity	★★★★
Landscape Value	★★★★
Wind Tolerance	★★★
Salt Tolerance	★★
Drought Tolerance	★★★
Flood Tolerance	★★
Cold Tolerance	★★★

The avocado is an important and familiar fruit crop. Once it is established, the tree is easy to maintain and produces an abundance of buttery, nutritious fruit. The richness in flavor correlates with a richness in calories. But the quality and utility of the fruit trump such concerns. Cold-hardy selections can be grown as far north as Gainesville. The tree is handsome and well behaved, although it can ultimately become quite large.

Known Hazards

The foliage is toxic.

GEOGRAPHIC DISTRIBUTION The avocado originated in Central America and the highlands of Mexico. Spanish settlers probably introduced the species into Florida at an early date. It was growing in the vicinity of St. Augustine by the early 1800s. Breeding efforts, conducted in Miami during the 1920s and 1930s, led to the development of many important cultivars. Commercial production in Florida is centered in Miami-Dade County.

TREE DESCRIPTION The avocado is a medium to large evergreen tree. Under ideal growing conditions, it may achieve a height of 70 feet. More typically, it attains a height of 30 to 40 feet. The canopy is moderately dense and may be spreading or pyramidal in form. The tree grows through tree-wide flushes of new leaves. Growth slows over the cooler months.

The deep-green leaves are alternate, smooth, and glossy. Generally elliptic in shape, they range from 4 to 13 inches in length. Emerging growth may have a reddish or brassy tinge. The root system is vigorous and can be invasive. Most roots are concentrated within the top 3 feet of soil. Feeder roots are heavily concentrated in a mat beneath the drip line of the canopy. The productive life of the tree is about 40 years.

Flower buds of a 'Day' avocado.

Dozens of tiny fruit may develop from a single flower panicle. The tree will shed all but one or two.

The emerging foliage of some cultivars is a brilliant scarlet.

FLOWERING AND POLLINATION The avocado may produce a million flowers during a single season. The yellow flower measures about 3/8 inch in diameter. Although it is perfect, the male and female parts are active at different times. Insects carry the pollen from tree to tree. In Florida, flowers may self-pollinate.

Flowering is a 2-day process. Trees are classified as Type A or Type B. Type A flowers open on the morning of the first day. During this period the stigma is receptive to pollen. They close on the afternoon of the first day. Type B flowers open on the afternoon of the first day, at which time the stigma is receptive to pollen. Both types remain closed overnight. On the second day, Type B flowers open first. The stamens shed their pollen during this opening. Type A flowers open in the afternoon of the second day, when the stamens shed their pollen. Pairing Type A and Type B avocadoes, though not always necessary, will ensure pollination and enhance fruit set.

FRUIT CHARACTERISTICS The fruit of the avocado is a large, specialized berry with a single seed. Fruit color ranges from green to black to reddish-purple. Fruit weight ranges from 4 ounces to more than 4 pounds. In shape the fruit may be pyriform, obovate, ellipsoid, or globose. The flesh is deep yellow or yellowish green. It is smooth in texture and buttery in flavor, often with a nutty component. The flesh contains between 3 and 28 percent oil. The seed fills an ovoid cavity in the center of the flesh. It is dense, astringent, and inedible.

RACES AND CULTIVARS Three races exist: the Guatemalan race, the Mexican race, and the West Indian race. Most cultivars grown in the United States are hybrids of 2 or more races.

The Mexican race originated in the Mexican highlands. The leaves give off a licorice scent when crushed. The fruit is small and pear shaped. Oil content is high. Flowering occurs in the winter and the fruit mature during the summer and fall. This race is salt intolerant. The Mexican race may survive brief temperature drops to 17° F. However, Mexican race avocados are poorly adapted to the heat and humidity of south Florida.

The Guatemalan race originated in the high-

Guatemalan race avocados often produce fruit with pebbled skin.

lands of Guatemala. Leaves are unscented. The fruit is medium to large and the skin has a pebbled texture. The fruit has medium oil content. Flowering occurs in the spring and the fruit mature during the fall and early winter. The tree is hardy to temperature drops of between 27° and 23° F. It is suitable for planting in south-central portions of the Florida peninsula.

The West Indian race originated in the tropical lowlands of Central America. Leaves are unscented. The fruit is medium to large and the skin is smooth. The seed is loosely embedded in the flesh. Oil content is low. Flowering occurs during late winter and the fruit mature over the summer. This race is moderately salt tolerant. However, it is vulnerable to cold temperatures, suffering serious damage at between 28° and 31° F.

Cultivars recommended for dooryard planting in north Florida and central Florida include:

'Brogdon' - This Mexican x West Indian hybrid is cold hardy, reasonably productive, and bears fruit of very good quality. The fruit, which average just over 10 ounces in weight, mature from July to November. The skin has a purple blush. Flowers are Type B. 'Brogdon' is hardy to about 22° F.

'Mexicola' - This Mexican-race cultivar bears small, obovate fruit, averaging less than 8 ounces. The skin is purplish-black. The seed is large in proportion to the fruit. Fruit mature from July through September. Flowers are Type A. The flesh is of very good quality. The tree will tolerate temperatures as low as 18° F.

'Winter Mexican' - 'Winter Mexican' is a Mexican x West Indian hybrid. Flowers are type B. The fruit are oval and are relatively small, weighing between 12 and 16 ounces. The skin is dark green and rough. The tree will tolerate temperatures as low as 22° F. Fruit mature from October through December.

'Gainesville' - This Mexican-race cultivar is extremely hardy. The fruit are small and green. The tree produces Type A flowers. It will survive temper-

ature drops to 17° or 18° F. Fruit quality is fair.

Cultivars recommended for south Florida and south central Florida include:

'Monroe' - This Guatemalan x West Indian cross is consistently productive and bears fruit of excellent quality. Fruit mature from late November through early February. Flowers are Type B. Fruit are medium, averaging between 18 and 20 ounces. 'Monroe' is hardy to about 26° F.

'Day' - 'Day' is a Guatemalan x West Indian hybrid that bears a fruit of very high quality. It produces Type A flowers. Fruit typically weigh between 10 to 16 ounces. The skin is green and somewhat dull. Fruit mature from July through September. The tree is cold hardy to about 24° F.

'Choquette' - 'Choquette' is a Guatemalan x West Indian hybrid that consistently bears large crops of high quality fruit. Flowers are Type A. Fruit mature from October through January. Fruit weigh between 25 and 38 ounces. 'Choquette' is hardy to about 28° F.

'Hall' - This Guatamalan x West Indian cross, produces superior fruit with yellow flesh. Flowers are Type B. Fruit are large, weighing between 20 and 30 ounces. Fruit mature over the winter. 'Hall' can endure a temperature drop to about 26° F.

'Lula' - 'Lula' is Guatemalan x Mexican hybrid that produces a good quality fruit and that can endure some cold. The fruit are small to medium, weighing between 10 and 20 ounces. Flowers are Type A. Fruit mature from October through December. The tree will survive a temperature drop to about 25° F.

'Black Prince' - This Guatemalan x West Indian hybrid is productive and yields a good quality fruit. The fruit are purple, averaging slightly over a pound in weight. They mature in August and September. Flowers are Type A. 'Black Prince' has endured temperatures to about 25° F.

RELATIVES The family Lauraceae contains about 30 genera and 2,000 species. The *Persea* genus contains about 150 species, mostly small trees native to tropical America and Southeast Asia. The genus *Cinnamomum* includes 3 species that are the source of the spice cinnamon. A number of Lauraceae species are native to Florida, including the red bay, *Persea borbonia,* and the swamp bay, *Persea palustris.* The leaves of these 2 species are spicy and aromatic and can be substituted for the leaves of the Mediterranean bay leaf tree in cooking. The sassafras, *Sassafras albidum,* is a distant relative of the avocado found in north Florida.

CLIMATE Hardy varieties can be grown as far north

as Gainesville and St. Augustine. Even in Jacksonville and the panhandle, the Mexican race avocado will grow, at least for a time, in protected locations.

CULTIVATION The avocado is a low-maintenance tree. Specimens have grown for many years without any care. The species is well adapted to most soil types as long as good drainage is present. The optimal soil pH is between 6.2 and 6.5, although the tree will endure a much broader range.

The tree is moderately drought tolerant. As a rule of thumb, it fairs better with too little water than too much water. Irrigation should be withheld if the soil is still damp from rainfall or from a previous watering. Where drainage is a problem, the tree can be planted atop mounds. Poor drainage is the number-one enemy of the avocado.

The tree is susceptible to damage from hurricane-force winds. One way to minimize wind damage is to top the tree at about 20 feet. In the Redlands area of Miami-Dade County, rows of sapodilla are often planted as windbreaks around avocado orchards.

The root system is sensitive and care should be taken not to break up the root ball when planting a tree. The grower should wait at least 2 months before fertilizing a newly planted tree. A mature specimen should be fertilized twice a year, once in late winter and once in early summer. The tree uses high levels of nitrogen during flowering and fruit development.

PESTS AND DISEASES Avocado scab is a fungal disease that appears as raised brown lesions on the skin of the fruit. It can be treated with approved sprays. Anthracnose, the bane of mango growers in south Florida, also affects the avocado. The infection begins as slightly sunken black spots, which spread across the surface of the fruit. Significant insect pests include the avocado lace bug, mites, scale insects, and red-banded thrips. Minor pests include the avocado looper, aphids, the avocado leafroller, mealybugs, and the banded cucumber beetle.

PROPAGATION Cleft grafting and veneer grafting are the primary methods of propagation. In regions susceptible to serious freezes, grafts are made close to the soil line. This allows the grower to protect the graft union by mounding soil and makes it likely that a damaged tree will regenerate from grafted wood. Seedlings generally produce fruit of inferior quality and may require more than 10 years to come into bearing.

HARVEST AND USE The fruit is harvested by hand or with clipping poles with attached canvas bags. In the home garden, the tree should be picked over on a weekly basis to produce a constant supply of fruit. The fruit is deemed physically mature when it reaches full size and after passage of a certain number of days from bloom. This number changes with the cultivar and location. Fruit from a particular cultivar matures over the same period each year. The mature fruit is still hard and will not ripen on the tree. It can therefore be stored on the tree for several weeks without any loss in quality. If the fruit is picked before it reaches maturity, it will not ripen properly. Once picked, the mature fruit soften and ripen in about a week.

These avocado trunks, looming out of the fog, represent a beginning rather than an end. The grove is being top worked to a superior cultivar.

Bael Fruit

SCIENTIFIC NAME: *Aegle marmelos*
FAMILY: Rutaceae

Fruiting Calendar

JAN	FEB	MAR	APR	MAY	JUN	JUL	AUG	SEP	OCT	NOV	DEC

Characteristics

Overall Rating	★★
Ease of Care	★★★
Taste/Quality	★★
Productivity	★★★
Landscape Value	★★★
Wind Tolerance	★★★
Salt Tolerance	★★
Drought Tolerance	★★★★
Flood Tolerance	★★★
Cold Tolerance	★★

The bael, a distant citrus relative, produces a hard-shelled fruit of minor import. While the fruit does not rank among the world's elite in terms of overall appeal, it has a pleasant taste and enjoys considerable popularity in India and elsewhere. The tree is easy to grow and fruits readily in peninsular Florida. It is moderately cold tolerant and can endure adverse growing conditions.

Known Hazards

The plant has sharp thorns capable of causing mechanical injury. The leaves and bark are thought to contain toxins.

GEOGRAPHIC DISTRIBUTION The bael originated in southern Asia, most probably India, although it has become naturalized in Pakistan, Sri Lanka, Bangladesh, Myanmar, and Thailand.

TREE DESCRIPTION The bael is a medium-size tree, typically reaching 30 to 40 feet in height and spread. It is reportedly deciduous in some regions, although it does not shed its leaves in south Florida. The crown is rounded and densely foliated. Young branches are armed with long spines. Leaves are alternate and are composed of up 3 or, occasionally, 5 leaflets. These are oval, toothed, medium green, and measure between 2 and 4 inches in length. Emerging growth may be tinted pink. The greenish-yellow, fragrant, bisexual flowers measure just under an inch in diameter. They occur in small clusters on new growth. The flowers have 4 petals and numerous stamens. The tree is relatively slow growing and typically comes into bearing between 5 and 7 years after it is planted.

FRUIT CHARACTERISTICS The fruit, which matures between 10 months and a year after bloom, is globular to ovoid in form. It usually measures between 3 and 6 inches in diameter. The green rind, which turns dull yellow at maturity, is thin, dense, and brittle. The rind contains yellow-orange to rust-colored flesh, pasty to somewhat gummy or fibrous in texture. The pulp is arranged in poorly defined segments radiating from a central core. The flavor is sweet, aromatic, and pleasant, although tangy and slightly astringent in some varieties. It resembles a marmalade made in part with citrus and in part with tamarind. Each fruit contains between 5 and 20 seeds, each measuring about 1/2 inch in length.

CULTIVARS Various selections have been made in regions where the bael fruit is popular. The cultivars 'Kaghzi' and 'Mitzapuri' are well regarded. Most Florida specimens are grown from seed.

RELATIVES The Rutaceae family, within the order Rutales and the subclass Rosidae, consists of about 150 genera and 900 species. It includes all *Citrus* species. This book presents separate discussions of the grapefruit, *Citrus* x *paradisi*; lemon, *Citrus limon*; lime, *Citrus aurantifolia*; mandarin, *Citrus reticulata*; and orange, *Citrus sinensis*. Non-*Citrus* species described within these pages include the kumquat, *Fortunella* spp., and white sapote, *Casimiroa edulis*.

CLIMATE The bael fruit can withstand some winter cold. It is reportedly capable of surviving temperatures as low as 19° F, placing it on a par with kumquats and cold-hardy forms of citrus. However, it is probably best suited to areas from Ocala southward. The tree prefers a monsoonal climate, with a pronounced dry period.

CULTIVATION The bael fruit is undemanding in its cultural requirements. It is drought tolerant. It is not particular as to soil, prospering on oolitic limestone, sand, clay, and other soil types with both high and low organic content. It will grow in soils with an extremely wide pH range. It is also said to be somewhat tolerant of poor drainage. The bael fruit is a tough tree and is not affected by any serious pests or diseases in Florida. It may occasionally harbor populations of mites, scale insects, and other pests that typically affect citrus. It is generally grown from seed in Florida. Efforts should be made to import superior varieties from India, Indonesia, and other growing regions. Budding is the preferred method of vegetative propagation.

HARVEST AND USE In Florida, the fruit ripens mainly over the summer. It should be harvested at color change, when the rind assumes a yellowish cast. In Florida, the average mature tree yields between 50 and 150 fruit per season. The fruit will store for 2 weeks at room temperature and for up to 2 months if refrigerated. The fruit may be cut in half and served as a breakfast fruit. It can also be dried or processed into drinks, jams, marmalades, and sauces.

The exterior of the bael fruit resembles that of its citrus cousins; however, the rind is a brittle shell.

Banana

SCIENTIFIC NAME: *Musa acuminata*
FAMILY: Muscaceae
OTHER COMMON NAME: Plátano (Spanish)

Fruiting Calendar

JAN	FEB	MAR	APR	MAY	JUN	JUL	AUG	SEP	OCT	NOV	DEC

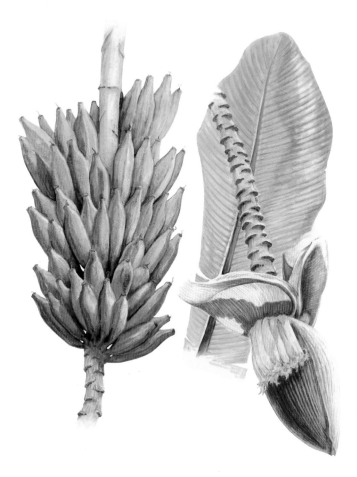

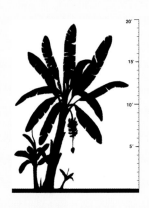

Characteristics

Overall Rating	★★★★★
Ease of Care	★★
Taste/Quality	★★★★★
Productivity	★★★★
Landscape Value	★★★
Wind Tolerance	★
Salt Tolerance	★
Drought Tolerance	★★
Flood Tolerance	★★★
Cold Tolerance	★★

The banana is a major fruit and is a staple food in many parts of the world. It shows astonishing variation in form, flavor, color, texture, and use. The Florida grower has access to unique cultivars that surpass store-bought bananas in flavor and appeal. The plant requires fertilization and irrigation. It is sensitive to cold and wind. Each shoot bears a single bunch then dies back—a process that may require 2 years. However, with proper management, the banana can be a productive and rewarding dooryard plant.

Known Hazards

None

36

GEOGRAPHIC DISTRIBUTION The cultivated banana originated in Southeast Asia, probably in the region of Malaysia. It has spread to nearly every tropical and subtropical region of the globe. The name banana is derived from the Arabic word for finger: banan. Small commercial plantings exist in Miami-Dade County, Palm Beach County, areas south of Lake Okeechobee, and Pine Island. The banana is a popular dooryard fruit throughout south Florida. Following a warm winter, the banana will fruit into central and north Florida.

PLANT DESCRIPTION The banana is not a tree, but is a large, herbacious perennial. What appears to be the trunk is actually a pseudostem composed of a series of stiff, tightly wrapped, concentric leaf sheaths. The "stem" of the banana is an underground rhizome, often referred to as the corm. This rhizome, which may live for several decades, develops buds from which the pseudostems form. In 9 to 18 months, the psuedostems reach their mature height of between 5 and 30 feet. The entire plant—the rhizome and the pseudostems—is known as a mat.

The Fruit and Spice Park in Miami-Dade County has an extensive collection of banana cultivars.

The large oblong leaves range from 2 to 9 feet in length. They are extensions of the sheaths of the pseudostem held on stout petioles. In warm weather the leaves form at the rate of one per week. The leaves may be green or may be variegated with white or red splotches. They are easily damaged in high winds.

FLOWERING The inflorescence grows up through the heart of the pseudostem. It forms once the plant has produced a certain number of leaves—usually between 35 and 45. The inflorescence holds rows of tubular white to yellow flowers. Female flowers emerge first. They are parthenocarpic and develop into fruit without the need for pollination. As the inflorescence continues to elongate, sterile flowers with male and female parts appear. Male flowers develop toward the tip. A fleshy reddish-purple bract protects each row of flowers. The bract opens, peels back, and is gradually shed as the fruit develop.

FRUIT CHARACTERISTICS The fruit of the banana is actually a specialized berry. Before harvest, the mature green fruit is very astringent. It contains 20 percent starch and only 1 percent sugar. Following harvest, the sugar content rises dramatically.

CULTIVARS Most modern banana cultivars descend from 2 wild species: *Musa acuminata* and *Musa balbisiana*. *Musa acuminata* lends its genes to modern dessert bananas. Most plantains or cooking bananas combine characteristics of *Musa acuminata* and *Musa balbisiana*. A few varieties, such as the Fe'i bananas of Polynesia, are derived from other species. Here is a list of dessert cultivars recommended for Florida.

'Blue Java' or 'Ice Cream' - This is a hardy, aggressive banana, well suited to the home garden. The plant typically achieves a height of 10 feet. It has fair cold resistance. The fruit is a silvery blue-green turning a washed-out yellow upon ripening. The fruit is 5 to 6 inches in length. The flesh is sweet, melting, and pure white. 'Blue Java' is vulnerable to Panama disease.

'Ice Cream' is one of the best bananas for dooryard planting in Florida.

'Goldfinger' (FHIA-1) - This variety, developed in Honduras, is very well suited to Florida. The fruit is delicious, mild, sub-acid and attains a length of 6 to 8 inches. The plant is disease resistant and moderately cold tolerant. The pseudostem attains a height of about 12 feet.

'Williams' - This popular cultivar produces a medium-size fruit with excellent flavor. It is resistant to Panama disease. The pseudostem grows to 8 to 10 feet. 'Williams' is somewhat tolerant of cold temperatures.

'Jamaican Red' - This banana comes in a dwarf version, which grows to about 8 feet, and a tall version, which grows to about 15 feet. The pseudostem and petioles are dark red. The fruit is delicious and turns from dark red to orange and yellow green upon ripening. The flesh is aromatic with cream to light orange coloration. 'Jamaican Red' is resistant to Panama disease, but is cold sensitive.

'Yangambi KM-5' - This is a sweet, top-quality, small banana, borne on a hardy, disease-resistant plant. It is also resistant to nematodes, which can be a serious problem in Florida's sandy soils.

'Cavendish' - This commercial variety is common in Florida. It comes in a dwarf and giant form. The plant is somewhat resistant to Panama disease. Although 'Cavendish' has some cold tolerance, leaves are damaged by near-freezing temperatures.

'Raja Puri' - This Indian cultivar produces a delicious, medium-size fruit. It is hardy and has some cold tolerance. However, it is susceptible to Black Sigatoka. The plant grows to 8 or 10 feet. It is moderately cold tolerant, although leaves are damaged when temperatures approach the freezing mark.

'Mysore' - This Indian cultivar, a *Musa acuminata* x *Musa balbisiana* hybrid, produces a small, very fine-tasting fruit. It is an important commercial crop in India. The fast-growing plant attains a height of about 15 feet. Mysore is resistant to Panama disease.

Other dessert-type cultivars suitable for growth in Florida include 'Aeae,' 'Brazilian,' 'Grand Naine,' 'Manzana,' 'Mona Lisa' ('FHIA-02'), 'Monkey Finger,' 'Nino,' 'Pisang Lemak Mani,' 'Pisang Raja,' 'SH 3640,' 'Super Dwarf,' and 'Viente Cohol.' Well-regarded cooking bananas include 'African Rhinohorn,' 'Balongkaue,' 'Bluggoe,' 'Cardaba,' 'French Horn,' 'Giant Plantain,' 'Orinoco,' 'Pelipita,' 'Saba,' and various Polynesian bananas, including 'Hua Moa,' 'Popoulu,' 'Maia Maoli,' 'White Iholena,' and 'Ha'a.'

RELATIVES Musaceae is a small family, composed of only 3 genera and 45 to 50 species. It falls within the monocotyledons group, which includes palms, grasses, and orchids. The cultivated banana comes from the genus *Musa*. Within the genus are such diverse species such as the baby pink banana,

The variegated banana 'Aeae' makes a stunning landscape specimen.

Musa velutina; the bloodleaf banana, *Musa zebrine;* the Indian dwarf banana, *Musa mannii;* the jungle banana, *Musa itinerans;* the ornamental banana, *Musa ornate;* the red torch banana, *Musa coccinea;* the scarlet banana, *Musa coccinea;* the wild forest banana, *Musa yunnanensis;* the Yunnan dwarf banana, *Musa rubinea;* and others.

Some species are purely ornamental and do not produce edible fruit.

CLIMATE Although tropical in habit, the banana is not confined to tropical, or even subtropical climates. In Florida, some varieties will grow and fruit as a seasonal crop as far north as the Georgia border. However, unless the plants are given a head start indoors, fruiting from Orlando north is sporadic. The optimal temperature range for growth lies between 78° and 90° F. Growth slows as the temperature drops and will shut down entirely when the temperature falls into the mid 50s. Meanwhile, temperatures exceeding 95° F also interfere with growth.

Most cultivars can endure a temperature drop to 30° F. A light frost may cause the foliage to burn but will usually not damage the pseudostem. On the other hand, temperatures of 28° or 29° F may kill the pseudostem back to the ground. This effectively ensures that the plant will not produce any fruit during the next season. However, unless a hard freeze occurs, the rhizome will generate new growth.

CULTIVATION The banana requires significant maintenance. Although the plant prefers deep loam, it grows reasonably well on sandy soil. The soil must be well drained. The banana will not tol-

The banana has poor wind resistance. Even moderate winds can shred leaves and uproot plants.

erate stagnant water or areas where the water table rises within a few inches of the surface. In low-lying areas, the plant should be established in a raised bed. The banana is intolerant of salinity in the soil or water. The optimal pH is between 6.0 and 6.5, although the plant will endure a much wider range. The banana is notoriously susceptible to wind damage. Winds above 50 miles per hour may topple and uproot plants. Protection can be achieved by providing a barrier hedge or by establishing plants in the lee of a dwelling or other structure.

The banana has some drought tolerance. However, irrigation will accelerate growth, stimulate fruit production, and cause the fruit to attain a larger size. During periods of high growth, some growers supply near-constant irrigation in the form of water pumped from a nearby pond or well. The plants seem to benefit from this treatment, so long as the soil is porous, the water is kept moving, and the irrigation is occasionally discontinued. Plants that receive irrigation and fertilization are far more productive than those that are neglected. A high-potassium fertilizer is recommended.

If too many pseudostems are permitted to grow from the same rhizome, fruit production will suffer. The pseudostems compete with each other for light and space, weakening the entire mat. Crowding can also lead to the emergence of pseudostems that lean at precarious angles. Unless excess pseudostems are culled, each successive pseudostem will put out a smaller bunch of fruit. The standard practice is to allow a maximum of 3 pseudostems to grow at any one time, cutting off any new growth as soon as it emerges. Newly formed pseudostems are referred to as pups or suckers. When the initial pseudostem reaches half of its ultimate height, one 'replacement' sucker is allowed to grow up beside it. Once the initial pseudostem puts out its inflorescence, a second 'replacement' sucker is allowed to come up. By rotating production in this manner, the grower may harvest 2 bunches from the same mat during a single growing season.

PESTS & DISEASES Black Sigatoka and Panama Disease are widely recognized as the most important diseases of the banana. The spread of these diseases—along with the fact that the banana is sterile—presents a long-term threat to production. The development of new, resistant hybrids continues. However, ultimately, genetic engineering may be required to save the banana from extinction. Nematodes harm the root systems of bananas in Florida and elsewhere.

PROPAGATION The cultivated banana does not produce seeds. The primary method of reproduction is through division; usually through separating and transplanting suckers. Suckers should be taken when they are between 2 and 3 feet tall. The banana can also be propagated from rhizomes that are dug up and carefully transplanted. Large rhizomes can be sectioned into several pieces, but each should have at least 1 or 2 buds.

HARVEST AND USE When intended for home use, the banana is harvested when the fruit begins to color. Bunches should be cut free with a machete or serrated knife. They should not be permitted to fall to the ground, as the weight of the bunch will crush and injure individual fruit. The bunch should be hung in a cool, shaded location. Individual hands may be removed as they ripen. The banana can be used for many purposes. It makes a superb additive to milkshakes, puddings, breads, ice cream, pies, and a host of other desserts. A cooking banana can be fried, baked, grilled, and used as an ingredient in various dishes.

Barbados Cherry

SCIENTIFIC NAME: *Malpighia punicifolia*
FAMILY: Malpighiaceae
OTHER COMMON NAME: West Indian Cherry, Acerola
(Spanish)

Fruiting Calendar

JAN	FEB	MAR	APR	MAY	JUN	JUL	AUG	SEP	OCT	NOV	DEC

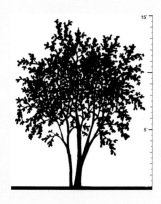

Characteristics

Overall Rating	★★
Ease of Care	★★★★
Taste/Quality	★
Productivity	★★★★
Landscape Value	★★★
Wind Tolerance	★★★★
Salt Tolerance	★★
Drought Tolerance	★★★
Flood Tolerance	★★★★★
Cold Tolerance	★★

The Barbados cherry is a minor fruiting species that produces a dependable crop in South Florida. The tree is compact, attractive, and requires little care. It is well adapted to Florida growing conditions and fruits abundantly over many months. As a dooryard tree it has many advantages. Better selections are mild and faintly sweet. However, the grumichama and cherry of the Rio Grande both produce a superior cherrylike fruit.

Known Hazards

None

GEOGRAPHIC DISTRIBUTION The Barbados cherry is thought to have originated in the southern islands of the Lesser Antilles or in adjacent areas of South America. It has long been a common tree throughout the Caribbean and Central America. It was first introduced into Florida in the 1880s.

TREE DESCRIPTION The Barbados cherry is a small evergreen tree. It may be trained as a tree or as a large shrub with multiple trunks. Without pruning, it will attain a height of about 15 feet. Some cultivars tend to be upright, while others tend to sprawl. The tree is densely branched. Leaves are shiny, deep green and vary in length from 2 to 3 inches. Leaves are opposite, pinnately veined, ovate to obovate in form. The root system is shallow, making the tree somewhat vulnerable to high winds. The growth rate is slow. However, the tree will usually begin to fruit within 2 years after being set out.

The Barbados cherry is an attractive landscape plant and lends color to the garden. It can be planted as a border hedge. Size can be readily controlled through judicious pruning. The pink flowers are small but attractive. They appear sporadically from spring through summer. An isolated plant may fail to set heavy crops as a result of pollination failure.

FRUIT CHARACTERISTICS The fruit is a medium-size, thin-skinned berry, with 3 indistinct lobes. The skin color ranges from fire engine red to deep crimson. As it matures, the fruit turns rapidly from green, to yellow, to red. The fruit is typically about an inch in diameter. It is borne on a pedicil originating from the leaf axil. The flesh is orange in color. The seed has 3 triangular, winged lobes. The unwieldy shape of the seed detracts somewhat from the eating quality of the fruit. The flavor ranges from tart and acidic to mildly sweet. The tree is highly productive, putting out as many as 4 or 5 crops a year.

CULTIVARS Several cultivars are regularly available from Florida nurseries. 'Florida Sweet' originated in Homestead. It forms an erect tree that produces large fruit that are mild and somewhat sweet. 'B-17' produces fruit of good quality. Other named selections include 'Beaumont,' 'ECHO sweet,' 'Haley,' 'Jamaican Giant,' 'Red Jumbo,' 'Manoa Sweet,' and 'Tropical Ruby.'

CLIMATE The Barbados cherry is sensitive to cold. It succeeds from Tampa and Cape Canaveral southward. An established tree may endure a temperature drop to 27° or 28° F. A young tree is more susceptible to cold injury.

RELATIVES The family Malpighiaceae consists of about 60 genera containing about 1,100 species. The Barbados cherry is the only species profiled within this book. The nance, *Byrsonima crassifolia,* is another notable species native to Central America and northern South America. The fruit is a yellow berry of good quality. The bunchosia or peanut butter fruit, *Bunchosia armeniaca,* is native to South America. This shrubby plant bears orange fruit with thick, sticky pulp.

CULTIVATION The Barbados cherry is a low-maintenance plant. It will tolerate some drought, although periodic irrigation leads to increased flowering and production. It will grow in a wide variety of soils, but requires proper drainage. Excessive fertilization may encourage vegetative growth at the expense of fruit production. When planted in sandy soil, the Barbados cherry is susceptible to damage from root-knot nematodes. Scales, aphids, and whiteflies occasionally attack the plant. The preferred method of propagation is through leafy hardwood cuttings.

HARVEST AND USE The fruit ripens 25 to 30 days after flowering. It is ready to pick when it attains full color. The Barbados cherry can be consumed fresh, can be juiced, or can be used in jams, jellies, and pie fillings. The fruit will not store for extended periods, but can be frozen.

The flower and immature fruit.

Blackberry

SCIENTIFIC NAME: *Rubus* spp.
FAMILY: Rosaceae
OTHER COMMON NAME: Zarzamora (Spanish)

Fruiting Calendar

JAN	FEB	MAR	APR	MAY	JUN	JUL	AUG	SEP	OCT	NOV	DEC

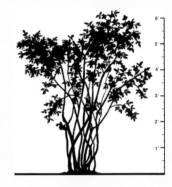

Characteristics

Overall Rating	★★★
Ease of Care	★★
Taste/Quality	★★★
Productivity	★★★
Landscape Value	★
Wind Tolerance	★★★★
Salt Tolerance	★
Drought Tolerance	★★★
Flood Tolerance	★★★
Cold Tolerance	★★★★★

The blackberry is easy to grow and produces an abundant, dessert-quality crop. While it is primarily a fruit of temperate climates, several varieties are well adapted to conditions in Florida. Although the fruit is perishable, this characteristic does not detract from its suitability for dooryard planting. The plant is rangy, aggressive, and thorny, making harvest difficult and placing some limits on its use within the home landscape.

Known Hazards

Sharp thorns are present on most varieties. The plants can spread unless soil barriers are installed.

GEOGRAPHIC DISTRIBUTION The blackberry is native to North America, including Florida. It will prosper in many habitats, but favors fence lines, roadsides, ditch banks, and overgrown clearings.

PLANT DESCRIPTION The blackberry is a sprawling bramble made up of multiple canes that emerge from the ground. Most varieties have thorns. The plant may be erect, trailing, or prostrate. It is not especially attractive from a landscaping perspective, although it can make a formidable hedge. The canes produce fruit during their second year of growth and die back thereafter. Individual plants may continue to produce new canes for 10 years. Leaves are compound consisting of 3 to 5 coarsely toothed leaflets. The white or pink flowers have 5 petals. They are borne terminally in loose clusters. Each flower contains between 40 and 80 tiny ovaries.

FRUIT CHARACTERISTICS The blackberry is an aggregate fruit and is not a true berry. It consists of as many as 80 druplets, each containing a tiny stone. The main characteristic that distinguishes the blackberry from the raspberry is the persistence of the fruit core or receptacle. In the blackberry, the receptacle remains attached to the fruit at harvest and is consumed with the fruit. In the raspberry, the drupes are pulled free of the receptacle.

SPECIES AND CULTIVARS The blackberry encompasses various species within the *Rubus* genus. Species native to Florida include the Florida prickly blackberry, *Rubus argutus;* the northern dewberry, *Rubus flagellaris;* the sand blackberry, *Rubus cuneifolius;* and the southern dewberry, *Rubus trivialis.* Several cultivars are suitable for dooryard planting in north and central Florida, including 'Kiowa,' 'Navaho,' 'Chickasaw,' 'Brazos,' 'Apachee,' 'Flordagrand,' 'Black Satin,' 'Oklawaha,' and 'Florida Seminole.'

RELATIVES The blackberry is a member of the large and diverse Rosaceae or Rose family. Besides the blackberry, this book features several member of the Rosaceae family, including apple, *Malus domestica;* capulin, *Prunus salicifolia;* chickasaw plum, *Prunus angustifolia;* loquat, *Eriobotrya japonica;* mayhaw, *Crataegus* spp.; peach, *Prunus persica;* pear, *Pyrus* spp.; and strawberry, *Fragaria ananassa.*

The raspberry, *Rubus* spp., is only marginally suited to cultivation in Florida. While the cultivar 'Dorman Red' can be grown in north Florida, the fruit is only of fair quality. The mysore raspberry, *Rubus niveus,* native to India and Myanmar, is recommended for south Florida. It produces small black fruit of good quality. However, it may be killed or severely damaged when the temperature falls to 26° F.

The mysore raspberry, a relative of the blackberry, is well suited to growth in south Florida.

CLIMATE The blackberry is rarely damaged by winter cold in Florida. Most cultivars have a winter chilling requirement of between 200 and 900 hours. A few are capable of producing adequate crops in central Florida. Wild blackberries grow in extreme south Florida, indicating that low-chill cultivars could be produced through cross breeding.

CULTIVATION The blackberry requires moderate maintenance. It is not particular as to soil type, but must have adequate drainage. Weed management is required to ensure productivity. Light applications of a balanced fertilizer, spaced at 2-month intervals from February through August, have been found to increase production. While the plant is drought tolerant, irrigation promotes fruit development. Erect varieties can be planted in hedgerows, spaced at 2- to 4-foot intervals. Trailing types require a trellis system.

The blackberry benefits from occasional pruning to remove dead canes and to prevent overcrowding. Canes may also be topped to promote branching. The plant spreads through underground runners, and new canes can emerge a considerable distance from the original patch. Because the blackberry can quickly overrun an area, the grower should consider burying a 12-inch, plastic soil barrier in a ring to contain the patch. Pests include birds, thrips, mites, and stinkbugs. The plant is amenable to various forms of vegetative propagation, including root cuttings and division.

HARVEST AND USE The fruit typically matures between 6 and 10 weeks after flowering, most often between May and August. It should be picked by hand when fully colored. Although it is excellent eaten out of hand, the fruit also makes an outstanding pie filling, jam, cider, and ice cream additive.

Black Sapote

SCIENTIFIC NAME: *Diospyros digyna*
FAMILY: Ebenaceae
OTHER COMMON NAMES: Chocolate Pudding Fruit,
    Zapote Negro (Spanish)

Fruiting Calendar

JAN	FEB	MAR	APR	MAY	JUN	JUL	AUG	SEP	OCT	NOV	DEC

Characteristics

Overall Rating	★★
Ease of Care	★★★
Taste/Quality	★★
Productivity	★★
Landscape Value	★★★
Wind Tolerance	★★★
Salt Tolerance	★★
Drought Tolerance	★★★
Flood Tolerance	★★★
Cold Tolerance	★

The black sapote—with its green skin and gooey dark-brown flesh—is little valued as a fresh fruit. However, it makes a superb dessert ingredient. Fruit mature over the winter when few other species come ripe. Drawbacks include low cold tolerance, messy fruit drop, and the large size ultimately attained by the tree. Despite these shortcomings, the black sapote makes a beautiful landscape specimen and has attained some popularity as a dooryard tree in south Florida.

Known Hazards
None

GEOGRAPHIC DISTRIBUTION The black sapote is native to the lowlands of Mexico and Central America. It is grown as dooryard crop in various tropical and subtropical regions. The species was first introduced into Florida in 1915, or thereabout. While it is not planted as a commercial crop, the fruit is occasionally sold from fruit stands in south Florida.

TREE DESCRIPTION The black sapote is a large, evergreen tree, sometimes attaining 80 feet in height. The canopy is thick, broad, and somewhat irregular. Leaves are alternate and oblong, measuring between 5 and 10 inches in length. They are deep green, glossy, and leathery. The tree makes an attractive landscape specimen and projects deep shade onto the area beneath the canopy. The tree has a moderate rate of growth. A young tree requires between 4 and 8 years before it will begin to bear fruit. Inconspicuous white flowers are borne in the leaf axils.

The leaves of the black sapote are glossy and ornamental.

FRUIT CHARACTERISTICS The fruit of the black sapote is oblate in form, somewhat resembling a green beefsteak tomato. The skin is light green and glossy, turning olive green upon maturity. The fruit typically measures between 3 and 5 inches in diameter. The pulp is nearly black, bitter, and very hard prior to ripening. It turns dark brown and becomes soft and custardlike when ripe. It is sweet, smooth, and creamy, with the taste and consistency of chocolate pudding. The fruit may be seedless or may contain up to 10 seeds.

CULTIVARS Few cultivars are available in Florida. In Australia, several cultivars have been selected, including: 'Bernecker,' 'Cocktail,' 'Cuevas,' 'Maher,' 'Mossman,' and 'Superb.'

RELATIVES Despite its name, the black sapote is not a member of the Sapotaceae or Sapote family. It belongs to the Ebenaceae or Ebony family. This small family is composed of 2 genera and about 400 species, most of which are trees and bushes with alternate, simple leaves. The black sapote is related to the persimmon, *Diospyro kaki,* and the mabolo, *Diospyros blancoi,* which are discussed separately in this book.

CLIMATE The black sapote is cold sensitive. Mature specimens suffer serious damage at about 28° F. Immature specimens may be killed at 30° F. Nevertheless, it succeeds as far north as West Palm Beach, on the east coast, and the Ft. Myers area on the west coast.

CULTIVATION The black sapote is a tree of low to moderate maintenance requirements. It will tolerate a wide variety of soil types, including clay, sand, and oolitic limestone. It can withstand brief flooding. The tree is said to benefit from occasional applications of fertilizer. However, it usually produces adequate crops without fertilization or irrigation and rarely suffers from nutritional deficiencies. The black sapote is usually grown from seed. Little variation exists among seedlings in terms of fruit quality. However, some trees have off-season bearing tendencies and some bear seedless or nearly seedless fruit.

HARVEST AND USE A mature tree produces more fruit than can be used by the typical family. In Florida, the fruit matures over the winter, from December through March. It is typically harvested with a cutting pole, when it has attained full size and undergone a slight color change. A mature fruit will ripen in 5 and 10 days at room temperature. The fruit is considered ripe when it becomes mushy and exhibits obvious signs of deterioration. Ripe fruit will store for about 4 days if refrigerated. The pulp is rarely eaten fresh. The flavor is enhanced by the addition of honey, vanilla, cream, or orange juice. The pulp can be used as a pie filling, in a mousse, or can be baked into breads. It can also be used as a flavoring for ice cream.

Although this fruit appears to be well past its prime, it is actually at the perfect stage for use in various recipes.

Blueberry

SCIENTIFIC NAME: *Vaccinium* spp.
FAMILY: Ericaceae
OTHER COMMON NAME: Arándano (Spanish)

Fruiting Calendar

JAN	FEB	MAR	APR	MAY	JUN	JUL	AUG	SEP	OCT	NOV	DEC

Characteristics

Overall Rating	★★★★
Ease of Care	★★★
Taste/Quality	★★★★
Productivity	★★★
Landscape Value	★★
Wind Tolerance	★★
Salt Tolerance	★
Drought Tolerance	★★★
Flood Tolerance	★★★★
Cold Tolerance	★★★★★

The blueberry is the quintessential North American berry and is a superb dooryard fruit. It bears an abundant crop on a handsome, compact bush. However, most varieties require exposure to weeks of cold temperatures to stimulate growth, flowering, and fruit set. The development of low-chill varieties has given residents of central Florida the opportunity to harvest a backyard blueberry crop. The fruit produced by these low-chill cultivars is surprisingly good.

Known Hazards

None

GEOGRAPHIC DISTRIBUTION The modern blueberry is a descendent of several species native to North America. Concentrated efforts at selection and breeding first occurred in the pine barrens of southern New Jersey during the 1910s. It was in this location that Dr. Fredrick Coville and Elizabeth C. White transformed the blueberry from a wild nibble into a commercial crop. Professor Ralph Sharp, of the University of Florida, is credited with breeding the modern low-chill blueberry. Starting around 1950, he selected specimens of the native lowbush blueberry from the vicinity of Winter Haven. He crossed these with rabbiteye and highbush varieties, and eventually managed to produce several cultivars suitable for growing in peninsular Florida. The United States accounts for nearly 90 percent of worldwide production. The blueberry is grown as a commercial crop in north and central Florida.

PLANT DESCRIPTION The blueberry is a deciduous, multi-caned bush. Although the plant can be low and sprawling, most cultivated varieties have an upright growth habit. Some types exceed 15 feet in height, although all can be kept in check through pruning. Depending on their ultimate size, bushes should be spaced at a distance of 4 or 5 feet.

Leaves are small, ovate to elliptic, measuring between 1 and 2 inches in length. The blueberry is shallow rooted and, therefore, should not be deeply tilled. New stems are wiry and bright green in color. Older growth has flaking grayish or reddish-gray bark. The blueberry is a plant of slow to moderate growth. Significant production does not take place until the plant reaches 3 or 4 years of age. The plant has a productive life of about 25 or 30 years, after which it experiences a gradual decline. A few specimens have born fruit over a period of 50 years.

The small, bell-shaped flowers are white to cream-colored. They are borne on new wood, forming dense clusters on short racemes. Most varieties benefit from cross-pollination. Some types will experience crop failures absent cross-pollination. Honeybees are important pollinating agents.

FRUIT CHARACTERISTICS The blueberry is borne in heavy clusters. The skin is thin, blue-black in color, but coated with a light, powdery bloom. Skin color changes from green to pink to red to blue as the fruit matures. The cream-colored flesh is medium-firm and contains a number of tiny seeds that are not noticeable during eating. Flavor ranges from tart to sweet.

SPECIES AND CULTIVARS Two varieties of blueberry are raised in Florida. The rabbiteye blueberry, *Vaccinium ashei,* forms a large bush and requires substantial chilling hours. It is well adapted to growth in north Florida. The southern highbush blueberry is a hybrid of *Vaccinium darrowi* and *Vaccinium ashei* or *Vaccinium corymbosum.* Southern highbush varieties form mid-size bushes. They tend to flower and fruit early. Southern highbush cultivars recommended for peninsular Florida include:

'Sharpblue' - This southern highbush cultivar, often simply referred to as 'Sharp,' was developed at the University of Florida. It is the most popular blueberry for peninsular Florida. Harvest occurs at the end of April or beginning of May. Although this variety is said to require 150 chilling hours, adequate fruit set has been observed with less than 50 chilling hours. The fruit is of good, though not outstanding, quality. The variety 'Misty' is often used as a pollinator, although 'Gulf Coast,' described below, seems to function as an adequate pollinator. 'Sharpblue' is not recommended for north Florida.

'Gulf Coast' - 'Gulf Coast' is a southern highbush cultivar released by the United States

These immature berries of the cultivar 'Sharp' will ripen in about two weeks.

Blueberry

Department of Agriculture in 1987. The fruit ripen around the first of May. It is said to require 200 to 300 chilling hours, but has been found to fruit adequately in areas that experience less than 100 chilling hours. While the fruit is of good quality, probably exceeding 'Sharpblue' in flavor, the stem and the dried calyx sometimes remain attached to the fruit. This defect requires extra steps at harvest.

'Emerald' - The University of Florida released this southern highbush cultivar in 1999. It has low chilling requirements but has not been thoroughly evaluated in southern parts of the peninsula. It reportedly ripens early, has a high yield, and produces a very large, good quality fruit.

'Misty' - 'Misty' is a low-chill, early-season variety with good fruit size and good quality for eating out of hand. However, this southern highbush cultivar varies in vigor and yield.

Cultivars recommended for north and north central Florida include 'Aliceblue,' a rabbiteye cultivar with low chilling requirements that ripens in early May; 'Austin,' a rabbiteye cultivar that produces high quality fruit in early June; 'Baldwin,' a mid- to late-season rabbiteye cultivar with heavy production; 'Biloxi,' an early-season, highbush cultivar, released by the USDA in 1998; 'Beckyblue,' a low-chill, early season rabbiteye cultivar released by the University of Florida in 1977; 'Bluebell,' a mid-season rabbiteye cultivar recommended for dooryard planting in north Florida; 'Bluecrisp,' a rabbit eye cultivar developed by the University of Florida that ripens in May; 'Bonita,' an early-season cultivar released by the University of Florida in 1985; 'Brightwell,' a productive rabbiteye cultivar developed by the University of Georgia that ripens in early June; 'Climax,' an early season rabbiteye cultivar developed by the University of Georgia; 'Georgia Gem,' a sweet-fruited, southern highbush cultivar that ripens in May; 'Millennia,' a southern highbush cultivar released in 2001 by the University of Florida; 'Ochlockonee' a late season cultivar developed by the University of Georgia and the United States Department of Agriculture; 'Premier,' a rabbiteye cultivar that ripens from late May through early June; 'Star,' a southern highbush variety released by the University of Florida in 1995 that ripens in May; 'Tifblue,' a high-yield, mid-season rabbiteye cultivar developed by the University of Georgia; and 'Windsor,' a southern highbush cultivar developed by the University of Florida that ripens from late April through early May.

RELATIVES The Ericaceae or Heath family encompasses about 100 genera and 2,000 species. The genus *Vaccinium* contains about 400 species.

The blueberry can be a lucrative crop in Florida, especially when raised in well-tended commercial operations.

Species native to Florida, include the highbush blueberry, *Vaccinium corymbosum,* of north Florida; Darrow's blueberry, *Vaccinium darrowii,* of northwestern and central Florida; the deerberry, *Vaccinium stamineum,* of north and central Florida; the shiny blueberry or evergreen blueberry, *Vaccinium myrsinites,* widely distributed throughout Florida; the small black blueberry, *Vaccinium tenellum,* of northeastern Florida; and the sparkleberry, *Vaccinium arboreum,* a large, treelike plant, widely distributed through north, central, and south central Florida. A familiar member of the *Vaccinium* genus that does not grow in Florida is the cranberry, *Vaccinium macroccarpus,* native to the bogs of northeastern North America.

The huckleberry produces a blueberry-like fruit laced with hard seeds. Huckleberry species native to Florida include the blue huckleberry, *Gaylussacia frondosa;* the dwarf huckleberry, *Gaylussacia dumosa;* and the woolly huckleberry, *Gaylussacia mosieri.* The wintergreen, *Gaultheria mucronata,* a low, woodland plant native to the northeastern United States, bears an aromatic, edible berry.

CLIMATE The blueberry is primarily a fruit of temperate climates. Some species grow north of the Arctic circle. When hardened off, the blueberry is not affected by the most severe low temperatures likely to occur in Florida. However, in north Florida, low-chill varieties sometimes suffer damage to flowers and new growth as a result of late frosts.

CULTIVATION The blueberry prefers acid soils with a pH of 4.0 to 5.2. Where the pH exceeds 6.5, the plant will exhibit low vigor and may become chlorotic. The plant grows especially well in sandy soils enriched with peat, sphagnum moss, and other organic matter. Mulching is important, particularly in the nutrient-poor soils covering much of the Florida peninsula. Competing weeds must be kept to a minimum.

The low-chill blueberry can bear very heavy crops, even in south central Florida.

The blueberry requires proper drainage and is often planted in raised beds. The plant requires weekly irrigation during periods of drought, particularly from March through May—the primary period of fruit development. The blueberry does not require irrigation over the winter. Excessive water during periods of dormancy can lead to root rot. The plant is sensitive to over-fertilization. However, it prospers with frequent, light applications of balanced fertilizer, a 15–10–10 formula, or some similar mix. Fertilizers formulated for other acid-loving plants such as camellias and azaleas work well.

The blueberry must be pruned to encourage proper development. Pruning should be conducted after harvest or during the dormant, winter period. The goal is to form a vase-shaped structure with between 6 and 10 dominant canes. Weak canes and spindly growth should be removed. Older canes should also be removed, as fruit production declines on canes more than 6 years old. Most authorities recommend the removal of flowers on plants less than 2 years old, to encourage vegetative growth.

PESTS AND DISEASES Several pests and diseases affect the blueberry in Florida. Scale insects, root weevils, leaf-footed bugs, stink bugs, thrips, and caterpillars cause minor damage. The blueberry gall midge feeds on flower and leaf buds and causes deformed leaves, stunted growth, and bloom failure. Birds peck V-shaped holes in the fruit. While the Caribbean fruit fly would seem to be a potential threat, infestation has been rare.

Root rot is a serious problem in areas with poor drainage. Flower blight may occur when rainy weather corresponds with periods of bloom. Both diseases can be controlled through the use of approved fungicides.

PROPAGATION The blueberry is most frequently propagated through hardwood cuttings. Whips, about the diameter of a pencil, are collected in February. These are divided into 6-inch cuttings, each containing one or more buds. The cuttings are placed upright in rooting medium, often composed of some mix of sand, sphagnum moss, and vermiculite. Softwood cuttings, taken shortly after harvest, may also be used. However, this technique requires use of a mist box. Grafting and budding are ineffective as the plants freely sucker below the graft union. A license is required to reproduce patented cultivars.

HARVEST AND USE A mature plant will produce 5 or more pounds of berries annually. The fruit does not ripen uniformly and bushes must be picked over several times. The blueberry can be consumed fresh or can be made into jelly, used as a filling for pies, added to muffins and breads, or used as flavoring in ice creams and milkshakes.

Few small fruit crops rival the blueberry in appearance or flavor.

Blue Grape

SCIENTIFIC NAME: *Myrciaria vexator*
FAMILY: Myrtaceae
OTHER COMMON NAME: False Jaboticaba

Fruiting Calendar

JAN	FEB	MAR	APR	MAY	JUN	JUL	AUG	SEP	OCT	NOV	DEC

Characteristics

Overall Rating	★★★
Ease of Care	★★★
Taste/Quality	★★★
Productivity	★★★
Landscape Value	★★★★★
Wind Tolerance	★★★
Salt Tolerance	★
Drought Tolerance	★★
Flood Tolerance	★★★
Cold Tolerance	★★

The blue grape is an exquisite landscape specimen that bears a delicious grapelike fruit. The species is obscure and is only available from specialized sources. Some consider the fruit slightly inferior to that of the jaboticaba, although it is of very good quality and pleasant flavor. As a result of its many attributes, the blue grape has excellent potential as a dooryard tree for south Florida.

Known Hazards

None

GEOGRAPHIC DISTRIBUTION The blue grape is native to South America, although its precise origin is uncertain. It has been grown in collections and scattered locations in south Florida for several decades. While the blue grape has never achieved significant popularity, this fact is attributable to its rarity and a general lack of familiarity rather than to any shortcoming associated with the tree.

TREE DESCRIPTION The blue grape is a small evergreen tree, reaching a height of 10 to 12 feet. It may have multiple trunks. It is usually heavily branched, beginning close to the ground. The rounded canopy, which extends in a skirtlike manner almost to the ground, is dense and finely textured. The bright green leaves, which measure from 3 to 5 inches in length, are opposite, entire, oblong-lanceolate, and pointed at the apex. They are glossy, slightly folded along the central axis, and point stiffly downward. The trunk is light gray. The bark, which is decorated with a reticulated pattern, occasionally peels in large patches, revealing a light undercoat. The tree is handsome, ornamental, and well behaved, and would be worth planting for its aesthetic value alone. Growth is very slow, averaging only about 10 inches per year. Flowering occurs in 2 or 3 waves, often in late winter or early spring. The small, white flowers appear both on older branches and on new growth. It is thought that cross-pollination is needed to ensure adequate fruit set.

The unique trunk and bark are representative of the overall beauty of the tree.

FRUIT CHARACTERISTICS The fruit of the blue grape is a thick-skinned berry, measuring from 1 to 1 1/2 inches in diameter. The skin changes from bright green to a dull bluish-purple as the fruit ripens. The rind is tough, rubbery, and inedible. The pulp is gelatinous and somewhat translucent. The color of the pulp is white tinged with pink or orange. The flavor is sweet and mildly aromatic.

Embedded within the pulp are 1 or 2 large, brown, kidney-shaped seeds. The ratio of edible flesh to waste is mildly disappointing.

RELATIVES The large and fruitful Myrtaceae or Myrtle family, within the order Myrtales, is thought to contain as many as 3,800 species. The genus *Myrciaria* includes the valuable jaboticaba, *Myrciaria* spp., reviewed within a separate section of this book. The camu-camu, *Myrciaria dubia;* false tamarisk, *Myrciaria borinquena;* rumberry, *Myrciaria floribunda;* and ridgetop guavaberry, *Myrciaria myrtifolia,* also fall within the genus. Other members of the Myrtaceae family, covered within this book, that share a distant affinity with the blue grape, include the cattley guava, *Psidium cattleianum;* cherry of the Rio Grande, *Eugenia aggregata;* feijoa, *Feijoa sellowiana;* grumichama, *Eugenia braziliensis;* guava, *Psidium guajava;* pitomba, *Eugenia luschnathiana;* and stoppers, *Eugenia* spp.

CLIMATE The blue grape's ability to tolerate winter cold has not been adequately assessed. It is not harmed by occasional light frosts in south Florida, although it may be somewhat less cold tolerant than its cousin, the jaboticaba.

CULTIVATION Maintenance requirements are low to moderate. The blue grape will grow on a wide variety of soils, including oolitic limestone and clay. It requires regular irrigation during establishment, but appears to have greater drought tolerance than the jaboticaba. The blue grape is relatively free of pests and diseases in Florida. The skin of the fruit is sufficiently tough to resist attack by the Caribbean fruit fly. The plant is almost universally grown from seed and little advantage would be gained through other forms of propagation. Seeds are slow to germinate, sometimes requiring 3 or more months to sprout.

HARVEST AND USE The blue grape is harvested when fully colored and slightly soft to the touch. The main crop matures in the spring, although fruit may be present throughout the warmer months. It is usually eaten out of hand, with the pulp sucked from the rind. The rind and seeds are discarded. The fruit spoils and dehydrates rapidly once it has been picked.

The blue grape is an exquisite landscape specimen.

Canistel

SCIENTIFIC NAME: *Pouteria campechiana*
FAMILY: Sapotaceae
OTHER COMMON NAMES: Eggfruit, Yellow Sapote, Fruta de Huevo (Spanish)

Fruiting Calendar

JAN	FEB	MAR	APR	MAY	JUN	JUL	AUG	SEP	OCT	NOV	DEC

Characteristics

Overall Rating	★★★★
Ease of Care	★★★★
Taste/Quality	★★★★
Productivity	★★★
Landscape Value	★★★
Wind Tolerance	★★★
Salt Tolerance	★★★
Drought Tolerance	★★★
Flood Tolerance	★★★
Cold Tolerance	★★

The canistel bears an outstanding fruit—showy, flavorful, nutritious, and useful. Yet the species is rarely seen outside tropical fruit collections. The tree is attractive and requires little maintenance. It comes into production over the winter when few other fruits are available. Admiration for this species grows with familiarity. It deserves wider planting and makes an ideal addition to the home landscape in south Florida.

Known Hazards

None

GEOGRAPHIC DISTRIBUTION The canistel origi-nated in southern Mexico and northern Central America. It is widely cultivated throughout Central America, the Caribbean, and the Bahamas.

TREE DESCRIPTION The canistel is a small to medium evergreen tree, attaining a height of about 30 feet in Florida. The foliage is dense, deep green, and glossy. Leaves are alternate, thickly clustered toward the branch tips. They are lanceolate or obo-vate, and measure from 5 to 10 inches in length. The tree drops most of its leaves in early spring at about the same time that new foliage begins to emerge. The small, bell-like flowers are bisexual and cream to light yellow in color.

FRUIT CHARACTERISTICS The fruit is highly vari-able. While typically ovate or spindle-shaped, with a broad base and pointed apex, it may be globose or oblate. The fruit usually measures from 2 to 5 inches in diameter. It may weigh from a few ounces to nearly 2 pounds. The ripe fruit is canary yellow, with smooth, glossy skin. The flesh is yellow and somewhat pasty in texture. In top-quality cultivars, the flesh is thick, rich, and creamy. The flavor is sweet and rich, suggesting an egg custard with a hint of pumpkin. The flesh contains 1 to 4 large brown seeds, each surrounded by a thin, transpar-ent membrane.

CULTIVARS Over the past several decades, growers have selected a number of promising strains, lead-ing to the emergence of a few superior cultivars. 'Trompo' (formerly '9681') was selected by the University of Florida's Tropical Research Extension Center in Homestead. The fruit is large, sweet, and of excellent texture 'Oro' (formerly '9680') was selected at the same location for its superior fruiting characteristics. 'Bruce,' produces large fruit of good quality. 'Ross Sapote,' a canistel-like fruit intro-duced from Costa Rica by rare-fruit pioneer, William Whitman, bears clusters of small, oblate fruits with moist flesh.

RELATIVES The canistel, a member of the fruit-rich Sapotaceae family, is distantly related to the sapodilla, *Manilkara zapote,* and star apple, *Chrysophyllum cainito,* and is more closely associ-ated with the abiu, *Pouteria caimito,* and mamey sapote, *Pouteria sapote,* all described within these pages. The lucmo, *Pouteria obbovata,* of the Andes region, produces a green-skinned fruit resembling a canistel, but of inferior quality. Its performance in Florida has been disappointing. The cinnamon apple, *Pouteria hypoglauca,* of Central America, bears a globose fruit, measuring between 2 1/2 and 4 inches in diameter. The flesh is sweet, slightly granular, and of pleasant flavor. This species grows

and fruits readily in south Florida. Other fruiting species within the genus *Pouteria* include the bully tree, *Pouteria multiflora,* of the Caribbean; the caimitillo, *Pouteria speciosa,* of the Amazon basin; the curiola, *Pouteria torta,* of Brazil; the fruteo, *Pouteria pariry,* of the Amazon basin; the green sapote, *Pouteria viridis,* of Central America; the lucma, *Pouteria macrophylla,* of the Amazon Basin; the maçaranduba, *Pouteria ramiflora,* of Brazil; and the níspero montañero, *Pouteria macrocarpa,* of Brazil.

CLIMATE The canistel requires a tropical or sub-tropical climate and is greatly harmed by tempera-tures below 28° F. In Florida, it grows as far north as Fort Pierce on the east coast, and Sarasota on the west coast.

CULTIVATION The canistel thrives with little cultural attention. It is not particular as to soil type. It does not require irrigation, except during the establishment of a young tree and in times of drought. A balanced fertilizer should be applied several times a year. The tree has few serious pests or diseases in Florida. Squirrels and opossums sometimes raid the tree for its fruit. The canistel grows readily from seed and can be cloned through cleft grafting, veneer graft-ing, budding, and air layering.

HARVEST AND USE In Florida, the fruit matures from November to March. It should be clipped from the tree leaving a small piece of stem attached. The fruit is best if picked when fully colored but still firm. A mature fruit will ripen within a week after it is harvested. The canistel can be eaten fresh. It excels when used as an additive to milkshakes or ice cream.

The 'Trompo' cultivar of the canistel, with its bright color and excellent flavor, is a fruit with commercial potential.

The seeds of the canistel, like those of most tropi-cal fruit species, are only viable for a short period once they have been removed from the fruit.

Capulin

SCIENTIFIC NAME: *Prunus salicifolia*
FAMILY: Rosaceae
OTHER COMMON NAME: Cereza Tropical (Spanish)

Fruiting Calendar

JAN	FEB	MAR	APR	MAY	JUN	JUL	AUG	SEP	OCT	NOV	DEC

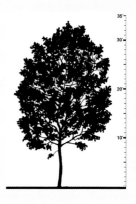

Characteristics

Overall Rating	★★
Ease of Care	★★★
Taste/Quality	★★
Productivity	★★★
Landscape Value	★★★
Wind Tolerance	★★★
Salt Tolerance	★
Drought Tolerance	★★★★
Flood Tolerance	★★★
Cold Tolerance	★★★

The capulin is a true cherry suitable for growth in subtropical climates. It is not related to other cherrylike fruit reviewed in this book, such as the Barbados cherry, cherry of the Rio Grande, Jamaica cherry, and grumichama. The fruit is exceeded in size, flavor, and quality, by fruit of the northern sweet cherry. However, select cultivars can bear fruit of good quality. The capulin deserves wider planting in Florida and is worthy of additional efforts aimed at improvement.

Known Hazards

The interior of the pit, as well as the twigs and foliage, are toxic.

54

GEOGRAPHIC DISTRIBUTION The capulin is native to the highlands of Mexico and Western Guatemala. It has been grown throughout Central America and Northern South America since before the arrival of the Spanish. The species has never achieved significant popularity in Florida, perhaps due to the fact that superior cultivars are rarely available.

TREE DESCRIPTION The capulin is a fast growing, medium-size tree. It attains a height of 40 feet or more in its native range. It is deciduous in cooler climates but sheds and replaces its leaves gradually in frost-free areas. Leaves are alternate, lanceolate, serrated, and measure from 4 to 6 inches in length.

The tree is upright in habit. The trunk is short. The bark is rough and gray. The wood is strong and dense; young branches are supple and tough. The capulin has good landscape value. The tree is vigorous and grows rapidly. White, fragrant flowers are formed on medium-length racemes from January to March.

FRUIT CHARACTERISTICS The fruit form pendulant clusters, containing 5 to 20 individual fruit. The fruit is round, measuring between 3/4 and 1 1/4 inches in diameter. The thin, glossy skin is usually deep red, purple, or nearly black. A few selections produce yellow-skinned fruit. The pulp is green, juicy, moderately sweet, and of pleasant, somewhat cherry-like flavor. It encloses a single hard stone. Areas adjacent to the skin may harbor a trace of astringency. Many rate the fruit to be nearly as good as that produced by the sweet cherry.

CULTIVARS Until recently, the capulin was exclusively reproduced from seed. Therefore, the capulin is still in its infancy in terms of improvement. Preferred cultivars include 'Ecuadorian,' 'Fausto,' 'Harriet,' 'Huachi Grande,' and 'La Roca Grande.' The quality of inferior varieties and seedlings is often disappointing.

RELATIVES The capulin is a member of the important and numerous Rosaceae family. Other species profiled within these pages include apple, *Malus domestica;* blackberry, *Rubus* spp.; chickasaw plum, *Prunus angustifolia;* loquat, *Eriobotrya japonica;* mayhaw, *Crataegus* spp.; peach, *Prunus persica;* pear, *Pyrus* spp.; and strawberry, *Fragaria ananassa.* Within the *Prunus* genus are such notable cultivated species as the apricot, *Prunus armeniaca;* the European plum, *Prunus domestica;* and the sweet cherry, *Prunus avium.* The northern sweet cherry requires exposure to substantial winter chilling and does not regularly fruit in Florida. Florida's native cherries, the black cherry, *Prunus serotin;* the Carolina laurelcherry, *Prunus*

Carolinian; and the threatened West Indian cherry, *Prunus myrtifolia,* produce small, tart fruit vastly inferior to those of the sweet cherry and capulin.

CLIMATE The capulin will grow in warm temperate or subtropical climates. It can withstand a temperature drop to 20° F, although twig damage will occur at 22° F and leaf damage will occur at 26° F. It is suitable for growth over most of the Florida peninsula.

CULTIVATION The capulin is a low-maintenance tree and is not exacting in its cultural requirements. It will tolerate a wide range of soils, including heavy clay and sand. It prefers a soil pH of 5.5 to 6.5. The capulin grows best in full sun. While drought tolerant, it benefits from irrigation during establishment and during fruit development. The tree also benefits from light applications of balanced fertilizer from spring through summer. Few pests and diseases affect the capulin in Florida. In south Florida, the Caribbean fruit fly sometimes invades the fruit. Mites and scales may require control. Birds occasionally raid the fruit. The capulin reproduces readily from seed, although seedlings bear fruit of variable quality. Selected cultivars can be reproduced using cleft grafts. Some success has also been obtained with semi-hardwood and softwood cuttings.

HARVEST AND USE In Florida, fruit typically ripen from May through and August, with production peaking in mid-summer. The fruit continue to improve if left on the tree for several days after reaching full color. The fruit can be used as a pie filling and may be substituted for the sweet cherry in any recipe that calls for the latter as an ingredient.

The fruit of Florida's native cherries can be used in preserves, but it is generally too tart to be eaten out of hand.

Carambola

SCIENTIFIC NAME: *Averrhoa carambola*
FAMILY: Oxalidaceae
OTHER COMMON NAME: Starfruit

Fruiting Calendar

JAN	FEB	MAR	APR	MAY	JUN	JUL	AUG	SEP	OCT	NOV	DEC

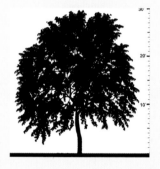

Characteristics

Overall Rating	★★★★
Ease of Care	★★★★
Taste/Quality	★★★
Productivity	★★★★★
Landscape Value	★★★
Wind Tolerance	★★★★
Salt Tolerance	★
Drought Tolerance	★★
Flood Tolerance	★★★
Cold Tolerance	★★★

No fruit tree provides greater near-term gratification than the carambola. It is precocious and bears abundantly over substantial portions of the year. The tree is handsome and is an appropriate size for most residential uses. The fruit should not be judged based on the sour, prematurely picked fruit sold by supermarkets. When harvested from a good cultivar at peak ripeness, the fruit is delicious: crisp, refreshing, and laden with sweet juice.

Known Hazards

None

56

GEOGRAPHIC DISTRIBUTION Some authorities place the origin of the carambola in Malaysia or other parts of Southeast Asia. Others assert that the carambola is native to Sri Lanka. The tree apparently no longer flourishes in its wild state. It has been cultivated in Southeast Asia, India, and southern China for many centuries. Today, the carambola is distributed throughout the tropics and is popular in Guiana, Brazil, the Philippines, and Australia.

In Florida, the carambola has been grown since the 1880s. Florida growers have actively participated in the improvement of this species through the evaluation and selection of superior cultivars. While several commercial operations exist, the carambola is most valued as a dooryard tree.

TREE DESCRIPTION The carambola is a medium-size, semi-deciduous tree. It commonly grows to 20 feet and occasionally reaches 35 feet. The canopy is densely foliated, rounded, and roughly symmetrical. Leaves are shed gradually toward the end of winter and replaced with new leaves so that the tree is never bare.

Leaves are alternate and compound, with 5 to 13 leaflets arranged in pairs. The leaflets measure from 1 to 3 inches in length. They are ovate, thin, with smooth margins. Leaflets are medium-green on the upper surface and light green below. They tend to fold at night or when the tree experiences drought-related stress. The trunk is short and covered with grayish-brown bark. The carambola has a medium rate of growth. Selected cultivars usually begin to bear fruit within a year after they are set out.

Flowers are born on short, branching panicles, which usually form in the axils of leaves, but sometimes emerge from mature wood. The flowers are small, purplish-pink, and attractive. They are about 1/4 inch in diameter, with 5 petals.

FRUIT CHARACTERISTICS The fruit of the carambola is an elliptic berry, measuring from 3 to 8 inches in length. It has 5 or 6 prominent longitudinal ribs. These form angles with one another, giving the fruit its characteristic star shape when sliced. The skin is thin, smooth, and slightly waxy. It is green at first, but turns yellow, orange, or greenish-white as the fruit ripens. The flesh is translucent and is of the same color as the skin. The flesh is very juicy, with the taste ranging from tart and flavorful to sweet and somewhat bland. Sweeter types are generally preferred in Florida. A narrow central core runs the length of the fruit. It may enclose up to 15 flat, brown seeds.

CULTIVARS Traits sought in a dooryard cultivar include high sugar content, thick skin, and sturdy,

The 'Bell' cultivar is capable of producing very large, sweet, high-quality fruit.

shallow ribs to minimize rubbing injury. Recommended cultivars are listed below.

'Kary' - This cultivar was selected in Hawaii. The medium-size fruit is very sweet, with thick ribs not prone to bruising. The skin is orange-yellow. 'Kary' bears prolific quantities of high-quality fruit.

'Sri Kembanqan' - This Thai cultivar is highly recommended for dooryard planting in Florida. It bears a slightly elongated fruit. The ridges sometimes remain green after the rest of the fruit ripens, producing a striped effect. The flesh is yellow-orange. It is sweet, with a delicious, rich flavor.

'Arkin' - 'Arkin,' formerly known as 'Star-king Sweetie,' is a leading commercial variety in Florida. It was selected in the late 1970s in Miami. The fruit is medium-large and sweet, with orange skin. It has excellent handling and shipping characteristics.

'B-17' - This orange-fruited Malaysian cultivar is sometimes referred to as the crystal honey carambola, in light of its high sugar content.

'Hart' - This white-skinned cultivar is well suited to dooryard planting. The fruit are small and delicate, but sweet and delicious, pearlike in flavor. The tree remains small and manageable.

'Bell' - 'Bell' bears large, yellow-skinned fruit of excellent flavor and quality. If thinned, the fruit can exceed 8 inches in length. However, the fruit are deeply winged and are susceptible to wind damage and rubbing injury.

'Fwang Tung' - This Thai variety bears large fruit with whitish-green skin. The flavor is mildly sweet. However, the fruit, like that of 'Bell,' is susceptible to damage because it is has thin, deeply cut ribs. Although it has poor shipping and handling characteristics, this does not diminish its value as a dooryard cultivar.

'Golden Star' - This quasi-tart variety originated in Homestead in the 1960s. The fruit is very attractive—large, flavorful, and deeply winged. The tree remains small and manageable. 'Golden Star' is a dependable and prolific bearer of good-quality fruit.

Other cultivars grown or tested in Florida include 'B-2,' 'B-10,' and 'B-16,' sweet-fruited Malaysian cultivars; 'Cheng Chui,' a sweet variety with light yellow skin; 'Hew-1,' a Thai selection that bears a sweet, good-quality fruit; 'Robert Newcomb,' a Florida selection bearing orange-yellow fruit with tart flavor; 'Thayer,' a tart variety selected in Florida; and 'Wheeler,' a Florida selection that bears orange fruit of good quality.

RELATIVES The carambola is a member of the Oxalidaceae or Wood-sorrel family. This small family consists of 7 or 8 genera and about 900 species.

The bilimbi, *Averrhoa bilimbi,* sometimes referred to as the cucumber tree, is a close relative of the carambola. This species may be native to the Molucca Islands. The tree attains a height of about 25 feet. The fruit, which is borne in great abundance, resembles a pickle or a bloated green carambola. The taste is sour and acidic. Unless the bilimbi is used as an addition to chutney or sauces requiring a tangy or sour element, it is usually combined with sugar to enhance palatability. The bilimbi is occasionally grown in south Florida, but is said to have greater cold sensitivity than the carambola.

The wood sorrels are distant relatives of the carambola. Several species are native to Florida, including the yellow wood sorrel, *Oxalis stricta,* a common plant throughout the state, and tufted yellow wood sorrel, *Oxalis lyonii,* native to North Florida. The wood sorrels are delicate, low-growing plants with cloverlike leaves. The leaves are edible and are sometimes used by hikers in salads or as a cooked vegetable.

CLIMATE The carambola has exhibited surprising tolerance of winter cold in Florida, enduring temperature drops to 25° F. The tree can survive in protected locations as far north as Clearwater and Daytona Beach. The severe 1989 freeze harmed but did not kill trees in Ft. Pierce on Florida's east coast.

CULTIVATION The carambola is a low-maintenance tree, requiring about the same amount of care as citrus. It will grow in a wide variety of soils, including sand and heavy clay. It prefers a soil pH of 5.5 to 6.5.

The tree is not tolerant of strong, prevailing winds. Leaves have a tendency to desiccate under such conditions. A tree planted in a wind-exposed location may be stunted and unproductive. With that said, the species is fairly resistant to hurricane-force winds. The leaves and fruit will tear away, but the tree itself usually remains intact.

Although the tree prefers full sun, it will tolerate some shade and still produce fruit. The carambola is moderately demanding when it comes to irrigation. Although it can survive short-term drought, productivity may suffer. Severe drought will cause the leaves to wilt and drop, and will cause immature fruit to drop prematurely. A newly planted tree requires regular irrigation until established. At the same time, the tree will not tolerate

The bilimbi, cousin of the carambola, bears a sour but useful fruit.

wet feet or constantly damp soil conditions. It is moderately tolerant of minor flooding of short duration.

When young, the carambola benefits from regular, light applications of balanced fertilizer. A mature tree should receive 2 applications annually: an application of 12–3–12 in the early spring and an application of 8–10–10 in the late summer. Chelated iron may be required to reverse chlorosis when the tree is grown in alkaline or limestone soils.

The young tree should be pruned to remove horizontal branches lower than 3 feet in height. The problem with low branches is that they tend to sag to the ground when loaded with fruit. As the tree matures, heading branches that protrude above the canopy will keep the size of the tree in check and will facilitate harvest. Efforts should be undertaken to maintain tree height at 15 feet or less.

Because production is prolific, the grower may need to thin excess fruit. A young treee may be doubled over with fruit and may be permanently disfigured unless the situation is corrected. Although the tree is pliant, limbs occasionally break under excessive fruit loads. Also, fruit are sometimes borne in tight clumps, causing rubbing injury.

PESTS AND DISEASES The carambola is not susceptible to any serious diseases in Florida. While fallen fruit may draw fruit flies and other insects, larvae of the Caribbean fruit fly rarely infest the ripening fruit, perhaps owing to the oxalic acid content. Scale insects sometimes attack the leaves and stems and may require occasional treatment. Stinkbugs sometimes cause stippling and skin discoloration. Squirrels generally ignore the fruit.

PROPAGATION The carambola is sometimes grown from seed, although the fruit borne by seedlings is erratic in quality. Seeds should be planted immediately after they are removed from the fruit. Seedling plants can often be collected from beneath the canopy of a mature tree. Various vegetative-propagation techniques are used. Budding, veneer grafting, and cleft grafting all have high rates of success. Air layering is rarely practiced due to slow root development.

HARVEST AND USE In Florida, multiple cropping is the rule. A mature tree will bear sporadically from July through February, sometimes setting 3 or 4 crops. Peak production occurs from September through January. A mature, well-irrigated tree may produce over 500 pounds of fruit. The fruit should be harvested when colored over at least 3/4 of its skin. Fruit picked prematurely will not develop full flavor. The carambola may be eaten out of hand. It is often cut into horizontal slices to achieve a decorative starlike effect. Some remove the outer edges of the ribs, although this is unnecessary. The fruit can also be made into jelly, used in salads, or juiced.

Although the flower is tiny, it is borne in great profusion.

Carissa

SCIENTIFIC NAME: *Carissa grandiflora* or *Carissa macro-carpa*
FAMILY: Apocynaceae
OTHER COMMON NAME: Natal Plum

Fruiting Calendar

JAN	FEB	MAR	APR	MAY	JUN	JUL	AUG	SEP	OCT	NOV	DEC

Characteristics

Overall Rating	★★★
Ease of Care	★★★★★
Taste/Quality	★★
Productivity	★★
Landscape Value	★★★★★
Wind Tolerance	★★★★
Salt Tolerance	★★★★
Drought Tolerance	★★★★
Flood Tolerance	★★★
Cold Tolerance	★★★

The carissa is widely grown in Florida as an ornamental hedge. The plant is tough, can tolerate unfavorable conditions, and will grow on barrier islands. The glossy, dark green foliage and beautiful white flowers provide visual appeal. The crimson fruit, which can be of very good quality, should not be permitted to go to waste. The carissa deserves recognition as an outstanding dual-purpose plant.

Known Hazards

The plant has very sharp spines. Some may find the latex mildly irritating.

GEOGRAPHIC DISTRIBUTION The carissa is native to South Africa. It has been grown in Florida since about 1886.

PLANT DESCRIPTION The carissa is a densely foliated, evergreen shrub. When left untended, the plant is sprawling in habit. It can be maintained in compact form, although pruning reduces fruit production. Leaves are simple and opposite, dark green, glossy, and leathery, measuring between 1 and 2 inches in length. They are ovate and entire.

The trunk is short and woody. Branches are tubular and flexible, often trailing to the ground. They are armed with prominent, forked thorns. The rate of growth is slow. The carissa can be used as a barrier hedge, foundation planting, or accent plant. The thorns make it inappropriate for planting in close proximity to play areas, walkways, or spigots. Flowers are white, showy, and fragrant, measuring between 1 and 2 inches in diameter. They have 5 waxy petals. The flowers appear at the branch tips over much of the year.

FRUIT CHARACTERISTICS The fruit is an oblong berry, measuring up to 2 1/2 inches in length. The apex may be somewhat pointed. The skin is crimson, thin, and tender. The flesh is pink, juicy, and melting. The flavor resembles that of a sweetened cranberry. In eating quality, it easily surpasses the Surinam cherry—another fruiting hedge plant common in south Florida. The small, disclike seeds, nestled in chambers near the center of the flesh, do not interfere with eating quality.

CULTIVARS In Florida, many cultivars have been selected for landscaping purposes. Many of these cultivars bear lightly, if at all. The selections 'Fancy,' 'Gifford,' and 'Torrey Pines' are said to bear fruit of good quality.

RELATIVES The Apocynaceae or Dogbane family is composed of between 200 and 360 genera and may contain as many as 3,700 species. A number of species are poisonous. Several, such as the oleander, *Nerium oleander,* and the be-still tree, *Cascabela thevetia,* produce beautiful flowers that belie the poisonous properties of the plant. About 40 species within the Apocynaceae family are native to Florida.

The *Carissa* genus contains about 35 species, mostly evergreen shrubs. The karanda, *Carissa carandas,* is native to southern Asia. It is grown to a limited extent in Florida. The white, 5-lobed flowers are borne in small clusters. The karanda produces a well-regarded purple fruit that measures about an inch in diameter. The num-num, *Carissa bispinosa,* is native to southern Africa. It resembles the carissa, but has smaller leaves and fruit. More distant fruit-ing relatives include the ceret muai, *Leuconotis eugeniifolius,* of Malaysia; the liane, *Landolphia buchananii,* of Africa; the pitabu, *Willughbeia angustifolia,* of Borneo; and the sorva, *Couma utilis,* of the Amazon region.

CLIMATE The carissa is subtropical in habit and will grow throughout most of the Florida peninsula. It succeeds as far north as Orlando and St. Augustine, although it is sometimes harmed by winter cold in these locations. An established plant may survive temperatures as low as 24° F.

CULTIVATION The carissa is a hardy, low-maintenance plant. It thrives in Florida's sandy soils and makes reasonable growth on the limestone soils of southeast Florida. The carissa benefits from periodic irrigation, although rainfall is usually adequate to sustain the plant. It is moderately salt tolerant, making it suitable for planting in coastal areas. The carissa is easily grown from seed. However, seedlings may be slow to come into production. The plant can also be propagated through air layering, ground layering, and stem tip cuttings.

HARVEST AND USE Fruit may be produced at any time of year, but production peaks just before Thanksgiving. Yield is moderate. The fruit is ripe when it turns a deep red. If the fruit is harvested too soon, the flesh will exude acrid white latex. The fruit may be eaten out of hand, made into preserves, or used as a pie filling.

The shape of the fruit varies greatly among cultivars.

The beautiful flower and foliage has made the carissa a popular choice for landscape purposes.

Cashew Apple

SCIENTIFIC NAME: *Anacardium occidentale*
FAMILY: Anacardiaceae
OTHER COMMON NAMES: Marañon or Anacardo
    (Spanish)

Fruiting Calendar

JAN	FEB	MAR	APR	MAY	JUN	JUL	AUG	SEP	OCT	NOV	DEC

Characteristics

Overall Rating	★★
Ease of Care	★★★
Taste/Quality	★★
Productivity	★★★
Landscape Value	★★★
Wind Tolerance	★★
Salt Tolerance	★
Drought Tolerance	★★★
Flood Tolerance	★★★
Cold Tolerance	★

From an economic standpoint, the cashew is the second most-important member of the family Anacardiaceae, following the mango. It produces both the familiar cashew nut, and the less well known cashew apple. The tree is attractive. The cashew nut is difficult to process and is not of interest to dooryard growers in Florida. The fruit, which forms around the base of the nut, is of fair quality and provides a juicy and refreshing treat. The tree is intolerant of winter cold and should not be attempted away from coastal areas of south Florida.

Known Hazards

Oils in nut case are toxic, caustic, and contain potent allergens. The leaves and sap may cause contact dermatitis in some individuals.

GEOGRAPHIC DISTRIBUTION The cashew is thought to have originated in eastern Brazil, but may have had a broader natural range. It has been widely planted throughout the tropics. Although it has been grown in south Florida for many years, it has never been common.

TREE DESCRIPTION The cashew is a small, spreading, evergreen tree. It can attain a height of about 30 feet. The tree may sprawl to encompass an equal or greater diameter. The leathery leaves are alternate, obovate, and measure from 5 to 8 inches in length. Emerging foliage is bright crimson. Older foliage may also take on various hues before it is shed. Small pink flowers, with 5 recurved petals, are born profusely on many-branched panicles. Each tree bears a mixture of male and female flowers. Growth is moderately fast, and the tree will ordinarily begin to bear within 3 years. Contact with the sap, foliage, or nutshell may cause dermatitis in sensitive individuals.

FRUIT CHARACTERISTICS The cashew apple is actually a pseudofruit and is composed, not of an ovary containing seeds, but of the swollen peduncle or receptacle immediately below the actual fruit. The true fruit is the nut. The nut develops first, and the peduncle then swells to surround the base. The cashew apple is pyriform in shape, measuring about 3 inches in length. The skin is thin and waxy, and changes from green to orange or bright red as the fruit ripens. The skin encloses a very juicy pulp laced with a fibrous, inedible rag. The flavor is subacid, slightly astringent, but nevertheless mild and pleasant.

The cashew nut protrudes from the apex of the cashew apple. It is nestled within in a double shell that contains caustic oil. In commercial operations, the nut is heat processed to make the shell brittle. If the shell is not properly removed from the nut, the toxic oil will contaminate the nut, rendering it inedible.

RELATIVES Other members of the Anacardiaceae family profiled within this book include the ambarella, *Spondias dulcis;* mango, *Mangifera indica;* and mombin, *Spondias* spp. The caja acu, *Anacardium giganteum,* is closely related to the cashew but forms a larger tree. It bears a pseudo-fruit with characteristics similar to those of the cashew.

CLIMATE The cashew requires a near-tropical climate. It has been grown as far north Palm Beach County along Florida's east coast. However, even brief exposure to freezing temperatures will burn leaves or cause defoliation. Temperatures below 29° F may be lethal.

CULTIVATION The cashew is a low-maintenance tree. It is drought tolerant and does not require irrigation once established. It will tolerate a wide range of soils, including oolitic limestone, sterile sand, and alkaline soils. However, it is not tolerant of salt in the soil or water. Regular fertilization increases the yield of nuts and, presumably, cashew apples. The cashew has few pests and suffers from few diseases in Florida. The main threat is posed by winter cold. The nut is slow to germinate and the germination rate is poor. Most trees are nevertheless grown from seed. While little advantage exists in vegetative propagation, the cashew may be reproduced through grafting and air layering.

HARVEST AND USE The cashew apple is borne over an extended period, from June through October. The fruit matures about 3 months after flowering. It is ripe when it separates easily from the stem or when it falls to the ground. The fruit is usually eaten out of hand. The interior rag is discarded. The fruit has poor storage characteristics and spoils rapidly at room temperature. It can be stored for a week or longer if refrigerated. The fruit can be candied or used in preserves. It can also be juiced, with the juice used as a beverage or flavoring.

The flowers and nut case of the cashew. The cashew apple swells toward the end of development.

The foliage can take on an array of colors.

Cattley Guava

SCIENTIFIC NAME: *Psidium cattleianum*
FAMILY: Myrtaceae
OTHER COMMON NAMES: Strawberry Guava, Guayabita
    Cereza (Spanish)

Fruiting Calendar

JAN	FEB	MAR	APR	MAY	JUN	JUL	AUG	SEP	OCT	NOV	DEC

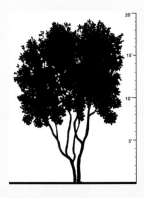

Characteristics

Overall Rating	★★
Ease of Care	★★★★★
Taste/Quality	★★★
Productivity	★★★★
Landscape Value	★★★★★
Wind Tolerance	★★★
Salt Tolerance	★★★
Drought Tolerance	★★★★
Flood Tolerance	★★★★★
Cold Tolerance	★★

The cattley guava is a familiar landscape plant in south Florida. The dark green foliage, mottled rust-colored bark, and shrublike growth habit combine to make it an eye-catching specimen. As a fruit tree, the cattley guava is inferior to its cousin, the common guava. The fruit is of good flavor and is borne in abundance, but is small and seedy. The Caribbean fruit fly frequently ruins the crop. The cattley guava is considered an invasive exotic.

Known Hazards

The plant is an invasive exotic.

GEOGRAPHIC DISTRIBUTION The cattley guava is thought to have originated in Brazil. It has been introduced into many regions of the tropics and subtropics. It has long been a popular landscape plant in peninsular Florida and is planted as far north as Tampa and Orlando. The cattley guava has escaped cultivation in several parts of the state, has formed thick stands, and has displaced native species.

TREE DESCRIPTION The cattley guava is an evergreen shrub or small tree. The leathery leaves are alternate, entire, obovate, and glossy. They typically measure between 1 1/2 and 3 inches in length. The flowers are typical of the Myrtaceae family, white, crowned by a prominent tuft of stamens. The cattley guava is self-compatible, it does not require cross-pollination to produce fruit.

FRUIT CHARACTERISTICS The fruit is a round to slightly obovoid berry, measuring about 1 1/4 inches in diameter. A persistent calyx is present at the apex. The central cavity contains numerous hard seeds surrounded by gelatinous pulp. The taste is aromatic, sweet, and somewhat similar to that of the common guava.

VARIETIES The species is actually made up of 2 distinct subspecies, the red or strawberry cattley guava, which attains a height of about 15 feet, and the yellow cattley guava, which grows to about 30 feet. The flesh of the strawberry cattley guava is pink, while that of the yellow cattley guava is light yellow.

RELATIVES Members of the Myrtaceae family that receive separate treatment within this book include the blue grape, *Myrciaria vexator;* cherry of the Rio Grande, *Eugenia aggregata;* feijoa, *Feijoa sellowiana;* grumichama, *Eugenia braziliensis;* guava, *Psidium guajava;* jaboticaba, *Myrciaria* spp.; pitomba, *Eugenia luschnanthiana;* and stoppers, *Eugenia* spp. The *Psidium* genus contains about 130 species, most of which are native to tropical America. Many produce edible fruit, including the araçá or para guava, *Psidium acutangulum;* Brazilian guava, *Psidium guineense;* Costa Rican guava, *Psidium friedrichsthalianum;* guayabita, *Psidium sartorianum;* Guinea guava, *Psidium guineense;* guayabo arrayán, *Psidium salutare;* Luquillo Mountain guava, *Psidium calyptranthoides;* mountain guava, *Psidium montanum;* and Puerto Rican guava, *Psidium microphyllum.* The mangroveberry or long-stalked stopper, *Psidium longipes,* is a rare plant native to south Florida. While not a member of the *Psidium* genus, the cambucá, *Plinias edulis,* is a guavalike member of the Myrtaceae family. It is native to the coastal rainforests of Brazil. This slow-growing tree bears 2-inch fruit with longitudinal ridges. The flesh is translucent, juicy, and of excellent flavor, said to resemble that of the papaya.

CLIMATE The cattley guava is one of the hardiest members of the *Psidium* genus and can withstand temperature drops to about 24° F. Even when frozen to the ground, the tree rapidly regenerates from the roots.

CULTIVATION The cattley guava is an extremely low-maintenance tree, requiring little or no cultural attention. It thrives in poor soil and can grow in poorly drained locations, considered too wet for most fruit trees. While the cattley guava is drought tolerant, fruit production may be poor unless the tree receives adequate water. The cattley guava is a primary host of the Caribbean fruit fly. In many years, this pest renders the fruit unfit for human consumption. At present, no solution exists to this problem. The fruit are simply too small and too numerous to bag on a regular basis. Infestation is less frequent in crops borne over the winter. The tree can be grown from seed or can be propagated by rooting leafy cuttings and air layering.

HARVEST AND USE The cattley guava is very productive and may bear 2 or 3 crops a year. The fruit is ripe when fully colored and when it separates from the tree with little pressure. The fruit deteriorates after 2 days at room temperature, but will last a week or more when refrigerated. The cattley guava can be eaten out of hand and can be made into pastes, purees, and jellies.

Developing fruit of the cattley guava. If fruit clusters are bagged at this stage, fruit fly infestation can be avoided.

Ceriman

SCIENTIFIC NAME: *Monstera deliciosa*
FAMILY: Araceae
OTHER COMMON NAMES: Swiss Cheese Plant, Monstera

Fruiting Calendar

JAN	FEB	MAR	APR	MAY	JUN	JUL	AUG	SEP	OCT	NOV	DEC

Characteristics

Overall Rating	★★
Ease of Care	★★★
Taste/Quality	★★★
Productivity	★★
Landscape Value	★★★★★
Wind Tolerance	★★★
Salt Tolerance	★★
Drought Tolerance	★★★
Flood Tolerance	★★★
Cold Tolerance	★★

The ceriman is an ornamental landscape plant with huge, dark green, perforated leaves. The fruit is peculiar in form, but appealing in taste. The species will endure considerable shade and can succeed as an understory plant, thus filling a unique niche within the garden. The plant is vigorous and requires little cultural attention. However, the ceriman is frost sensitive and can become rampant in its habit of growth.

Known Hazards

All parts of the plant are poisonous. The unripe fruit contains irritating calcium oxalate crystals.

GEOGRAPHIC DISTRIBUTION The ceriman is native to the rain forests of Central America and southern Mexico. It has been distributed throughout warm regions of the world, primarily as an ornamental. Perhaps surprisingly, it has been the subject of small-scale commercial operations in south Florida.

PLANT DESCRIPTION The ceriman is a stout, herbaceous vine, similar in appearance to its cousin, the split-leaf philodendron. The plant may grow horizontally, attaining a length of more than 50 feet, or may climb to a height of more than 30 feet. The stems are cylindrical and solid, measuring about 3 inches in thickness. They produce thick mats of clinging, cordlike roots. Leaves are large, glossy, deep green, and leathery, measuring up to 3 feet in diameter. They are perforated with oblong holes. The margins are deeply notched. The leaves are borne on stiff petioles. These emerge in dense whorls from the growing tip. As the leaves age and drop off, they leave persistent scars, spiraling about the stem. The ceriman excels as an accent plant or as a foundation plant. It will also grow underneath other fruit trees, making double use of valuable space. Flowers are borne on spathes supported by long stalks that emerge from the leaf axils. The tiny flowers are perfect, containing both male and female parts. Flowers and ripening fruit may be present on the plant at the same time.

FRUIT CHARACTERISTICS The fruit of the ceriman is compound. It typically measures about 10 inches in length and 2 1/2 inches in width. The rind is made up of hexagonal, tilelike scales, which cover segments of pulp. The pulp is off-white, crisp, juicy, and sweet. It is flecked with black floral remnants. In flavor, the ceriman resembles a pineapple.

CULTIVARS Named cultivars of ceriman are rare. A few, such as 'Albovariegata' and 'Variegata,' have been selected based on their ornamental characteristics.

RELATIVES The ceriman is the only noteworthy fruiting plant within the family Araceae. This large family contains about 2,000 species in 115 genera. Most are herbs, vines, or climbing shrubs. The rhizomes of some species are used as starchy vegetables, although most are caustic and inedible prior to cooking. Perhaps the best known of these is the taro, *Colocasia esculenta,* from southern Asia. Other Araceae species include the philodendrons, *Philodendron* spp.; elephant ear, *Xanthosoma sagittifolium;* and titan arum, *Amorphophallus titanium,* said to produce the world's largest flower. Several members of the family are native to Florida, including the goldenclub, *Orontium aquaticum;* Jack-in-the-pulpit, *Arisaema triphyllum;* and white arrow arum, *Peltandra sagittifolia.*

CLIMATE The ceriman is killed by a brief exposure to freezing temperatures. It is therefore relegated to southern coastal areas of the state.

CULTIVATION The ceriman requires little attention. It is one of only a handful of fruiting plants that do well in the shade. Although it prefers well-drained soil, it will grow in soil that is moist and occasionally saturated. Fertilization may speed growth but is usually unnecessary. Few pests bother the plant. The lubber grasshopper has been known to attack the foliage. Additional minor pests include leaf spots, scales, and mealybugs. The ceriman is propagated by means of stem cuttings.

HARVEST AND USE The fruit ripens about a year after bloom. As it ripens the rind lightens and segments near the base begin to separate. The fruit is clipped from the petiole about an inch from the base. After harvest, the fruit continues to ripen slowly over a period of several days, with individual segments coming ripe from the base toward the apex. Segments are ripe when the rind begins to peel from the underlying flesh and can be removed with minimal effort. Only fully ripe segments should be consumed, as immature segments contain needlelike crystals of calcium oxalate, which can cause serious irritation to the mouth and throat. All parts of the plant other than the fruit are poisonous.

The ceriman is one of only a handful of fruit plants that are productive in shady locations.

Che

SCIENTIFIC NAME: *Cudrania tricuspidata*
FAMILY: Moraceae
OTHER COMMON NAMES: Chinese Mulberry, Mandarin
   Melon Berry

Fruiting Calendar

JAN	FEB	MAR	APR	MAY	JUN	JUL	AUG	SEP	OCT	NOV	DEC

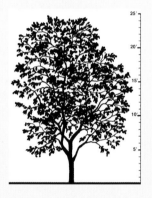

Characteristics

Overall Rating	★★★
Ease of Care	★★★★
Taste/Quality	★★★
Productivity	★★★
Landscape Value	★★★
Wind Tolerance	★★★
Salt Tolerance	★
Drought Tolerance	★★★
Flood Tolerance	★★
Cold Tolerance	★★★★★

Although it has not yet achieved the popularity it deserves, the che is well suited to dooryard planting in north and central Florida. This small tree is easy to grow, cold hardy, drought resistant, and produces a berrylike fruit in great abundance. The flavor is mild, sweet, and delicious. No significant faults are associated with the che, apart from sharp thorns and the tendency of some juvenile trees to form suckers around the base.

Known Hazards
Terminal branches may be armed with sharp thorns.

68

GEOGRAPHIC DISTRIBUTION This species is native to China and East Asia. It is widely distributed in southern Europe and other warm temperate regions of the globe.

TREE DESCRIPTION The che is a small, deciduous tree or spreading bush. When trained as a tree, it may attain a height of 20 feet or more. Trees grown from cuttings tend to sucker, forming dense bushes. Leaves are alternate, often with 3 indistinct lobes, although the leaves show great variability in form. They typically measure 3 to 5 inches in length. Most branches are armed with sharp thorns. The che should not be planted over walks, drives or patios, as fallen fruit tends to stain porous surfaces.

The che is dioecious: some trees bear only male flowers while others bear only female flowers. The extent to which both types of trees are needed for fruit set is unsettled. A female tree may bear fruit without any apparent nearby male pollinator. Flowers appear in the spring. They are small, forming round, greenish yellow heads that usually measure less than 1/2 inch in diameter.

FRUIT CHARACTERISTICS The che is a collective fruit. It outwardly resembles a fused raspberry. The fruit measures 1 to 1 2/3 inches in diameter. As it ripens, the fruit turns from green to yellow to red. In mature fruit, both the skin and interior flesh are red. In taste the che resembles a cantaloupe crossed with a fig. Each fruit contains a few tiny brown seeds.

CULTIVARS Few named cultivars are available in the United States. 'Darrow' is sometimes planted in north Florida.

RELATIVES The Moraceae or Mulberry family, within the order Urticales and the subclass Hamamelidae, includes about 60 genera and 2,000 species. Separate sections of this book are devoted to 3 other members of the Moraceae family: the fig, *Ficus carica;* the jackfruit, *Artocarpus heterophyllus;* and the mulberry, *Morus* spp. The unusual Osage orange, *Maclura pomifera,* is a member of the Moraceae family native to the south-central United States. It is sometimes used as a rootstock for the che. The softball-sized fruit exudes a pleasant, citruslike odor. However, the flesh is tough and inedible.

CLIMATE The che can withstand temperature drops to -20° F and is hardy in all regions of Florida. It is probably somewhat better suited to north Florida than south Florida, although its performance in southern parts of the peninsula has not been adequately assessed.

CULTIVATION The che is a hardy, low-maintenance tree. It prefers sandy, mildly acidic soil, but will tolerate a wide range of soil types. A young tree should be given periodic irrigation until established. Once established, the che is drought tolerant and rarely requires supplemental irrigation. It is not tolerant of flooding. The tree should be pruned during winter dormancy by heading back year-old branches. This helps stimulate new growth and fruit set.

This species does not appear to be attacked by any serious pests or diseases in Florida. Birds and various mammals may raid the tree, but production is usually sufficient to offset any losses. The che may be grown from cuttings or from seed. Softwood cuttings are taken in mid-summer and are kept under mist or planted in moist sand following treatment with a rooting hormone.

HARVEST AND USE Fruit matures from July to September. A mature tree may yield in excess of 200 pounds. The fruit should be permitted to achieve full coloration prior to harvest. It does not achieve ideal flavor and sweetness until it turns a dark maroon-red. Ripeness is also indicated by a slight softening and by the absence of latex when the fruit is pulled from the tree. At room temperature, the che has a shelf life of about 3 days. It will store for about a week in the refrigerator. The fruit is most often eaten out of hand, but can be substituted for blackberries, figs, raspberries, or mulberries in various recipes.

The immature fruit of the che are tiny; however, the fruit develop rapidly as they approach maturity.

Cherapu

SCIENTIFIC NAME: *Garcinia prainiana*
FAMILY: Cluciaceae
OTHER COMMON NAME: Button Mangosteen

Fruiting Calendar

JAN	FEB	MAR	APR	MAY	JUN	JUL	AUG	SEP	OCT	NOV	DEC

Characteristics

Overall Rating	★★★
Ease of Care	★
Taste/Quality	★★★★
Productivity	★★★
Landscape Value	★★★★
Wind Tolerance	★★
Salt Tolerance	★★
Drought Tolerance	★★
Flood Tolerance	★★
Cold Tolerance	★

The cherapu is the most cold-sensitive plant profiled in this book. When planted outdoors, it is likely to succumb to winter cold in all locations north of the Keys. The species is nevertheless suitable for growing as a container plant and can be moved indoors when frost threatens. It produces an outstanding fruit, approximating the flavor of its revered but ultra-tropical cousin, the mangosteen. For Florida growers who yearn for fresh mangosteen, the cherapu presents a respectable alternative.

Known Hazards

None

GEOGRAPHIC DISTRIBUTION The cherapu is thought to be native to Thailand or Malaysia, but is widely grown throughout Southeast Asia.

TREE DESCRIPTION The cherapu forms a small to medium evergreen tree. In its native habitat it can reach a height of 20 feet or more. However, it is slow growing and acclimates readily to growth in a moderate-size container. It can easily be maintained at a height of between 3 and 6 feet. Fruiting takes place when the plant attains a height of about 3 feet. The glossy, dark-green leaves are opposite, ovate-oblong, and measure between 2 and 5 inches in length. The 5-petaled, red flowers are small but showy. The cherapu is dioecious; it requires both a male and female tree to ensure fruit set. Male flowers are nearly identical to female flowers, but display a ring of short stamens coated with yellow pollen. Hand pollination, accomplished by touching a male bloom to a female bloom, nearly always results in fruit set.

FRUIT CHARACTERISTICS The fruit, round in shape, measures about 1 1/2 inches in diameter. The skin is smooth, green when immature, turning a rich orange upon reaching maturity. A lobed calyx is attached to the base. The apex is decorated by a raised, brown 'button' made up of stigma remnants. The thin rind contains between 4 and 8 segments of fragrant, orange flesh of pleasant, subacid flavor.

RELATIVES The Clusiaceae or Guttiferae family, commonly known as the Garcina family, exists within the order Theales and subclass Dilleniidae. It is thought to contain about 1,100 species within 50 genera. The *Garcinia* genus is made up of approximately 200 species. The most renowned member of the genus is the mangosteen, *Garcinia mangostana*. This medium-size tree is native to Southeast Asia. The purple fruit contains between 4 and 8 segments of white flesh of superlative flavor. Temperatures below 40° F damage this ultra tropical species. Other fruiting *Garcinias* include the Brunei cherry, *Garcinia parvifolia*; cherry mangosteen, *Garcinia intermedia*; gamboge, *Garcinia xanthochymus*; gourka, *Garcina dulces*; and seaside mangosteen, *Garcinia hombroniana*. Besides the cherapu, the only member of the Clusiacea family covered within this book is the imbe, *Garcinia livingstonei*. Several New World *Garcinias*, formerly classified within the genus *Rheedia*, are listed among relatives of the imbe.

CLIMATE The cherapu is greatly harmed at temperatures below 36° F. Therefore, it is not suitable for growth in peninsular Florida unless sheltered in a greenhouse or raised as a container specimen.

CULTIVATION The cherapu is a species for the dedicated enthusiast. At a minimum, the plant requires adequate cold protection, hand pollination (the presence of a male plant), shelter from strong winds, and regular watering. It is not tolerant of drought. A young tree should be protected by shade cloth until it achieves a height of about 2 feet. The cherapu appears to prefer slightly acidic soil with some organic content. A complete, slow-release fertilizer should be applied at 3- or 4-month intervals. Some growers graft a male branch onto a female tree to aid in pollination. Few pests or diseases affect the cherapu in Florida. Red-banded thrips are an occasional problem. The cherapu is usually grown from seed. Seeds should be planted within a day after they are removed from the fruit. They usually germinate within 2 months, although delayed germination is common.

HARVEST AND USE Fruiting usually takes place during the summer but can occur continuously over the warmer months. The fruit should be clipped from the branch when fully colored, leaving a short stub of stem protruding from the calyx. It is best eaten out of hand. The bitter rind should be discarded. The fruit can be stored for about a week at room temperature and can be stored for up to 2 weeks in the refrigerator.

The female flower of the cherapu lacks the ring of stamens that is present in the male (shown in illustration on previous page).

Cherry of the Rio Grande

SCIENTIFIC NAME: *Eugenia aggregata*
FAMILY: Myrtaceae

Fruiting Calendar

JAN	FEB	MAR	APR	MAY	JUN	JUL	AUG	SEP	OCT	NOV	DEC

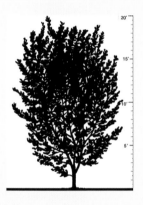

Characteristics

Overall Rating	★★★
Ease of Care	★★★
Taste/Quality	★★★
Productivity	★★★
Landscape Value	★★★
Wind Tolerance	★★★
Salt Tolerance	★
Drought Tolerance	★★★★
Flood Tolerance	★★★
Cold Tolerance	★★★

The cherry of the Rio Grande produces an excellent, cherrylike fruit, which is universally regarded as among the best within the *Eugenia* genus. The tree is compact, attractive, vigorous, and relatively cold hardy. However, the tree is susceptible to a mysterious dieback that is rarely fatal, but that may be disfiguring. Maggots of the Caribbean fruit fly occasionally attack the fruit.

Known Hazards

None

72

GEOGRAPHIC DISTRIBUTION The cherry of the Rio Grande is native to Brazil, specifically the state of Rio Grande do Sul. It has been grown in Florida since prior to 1920, but has never achieved widespread popularity.

TREE DESCRIPTION The cherry of the Rio Grande is an evergreen shrub or small, compact tree. While it reportedly grows to 30 feet in Brazil, in Florida the tree rarely exceeds 15 feet. The tree is heavily branched, forming a dense canopy. Leaves are glossy, deep green with smooth margins. They are narrow elliptic, often slightly folded, and measure 2 to 3 inches in length. The tree has a slow to moderate rate of growth. Seedlings require 4 or 5 years to bear fruit.

Like many *Eugenias*, the cherry of the Rio Grande makes an attractive landscape specimen. It is sometimes used as a hedge. Because of its relatively small size, it can be maintained for many years as a container specimen. The tree typically begins to flower in late February or early March. Flowering continues for several months, sometimes extending into May. Flowers are small and white but showy.

FRUIT CHARACTERISTICS The fruit of the cherry of the Rio Grande is deep red in color and oblong in form, ranging from 3/4 of an inch to 1 1/4 inches in length. The taste is sweet and pleasant. Of all tropical fruitspecies—perhaps with the exception of better cultivars of Capulin—the cherry of the Rio Grande produces a fruit most similar to that of a sweet cherry. The fruit is superior in flavor to that of the Surinam cherry and is comparable to that of the grumichama—which is to say that it is very good.

CULTIVARS Several unnamed selections are available in Florida.

RELATIVES The *Eugenia* genus of the Myrtaceae family is made up of about 400 species. Besides the cherry of the Rio Grande, 3 *Eugenias* are widely planted for their fruit. They are the pitomba, *Eugenia luschnanthiana;* the Surinam cherry, *Eugenia uniflora;* and the grumichama, *Eugenia braziliensis*. Other fruiting members of the *Eugenia* genus, native to South America, include the araçá-boi, *Eugenia stipitata;* cabeludinha, *Eugenia tomentosa;* pêra-do-campo, *Eugenia klotzchiana;* perinha, *Eugenia lutescens;* and uvaia, *Eugenia pyriformi*. More distant relatives of the cherry of the Rio Grande, covered within this book, include the blue grape, *Myrciaria vexator;* cattley guava, *Psidium cattleianum;* feijoa, *Feijoa sellowiana;* guava, *Psidium guajava;* and jaboticaba, *Myrciaria* spp.

CLIMATE Although the species is subtropical in habit, it is surprisingly hardy. A mature specimen may survive a temperature drop to 22° or even 20° F. The tree grows in Orlando and Tampa and succeeds as far north as St. Augustine on Florida's east coast.

CULTIVATION The cherry of the Rio Grande is a low-maintenance plant and will prosper in most soils. A young tree should be periodically irrigated until established. A mature tree may require irrigation during drought and during fruit development. Palm fertilizer has been found to speed growth and development. The cherry of the Rio Grande is a host of the Caribbean fruit fly. It also suffers from limb dieback of unknown cause. The disease is common and weighs against the plant's use as a landscape specimen. Infected limbs should be removed. The cherry of the Rio Grande is usually grown from seed. Seeds typically germinate in less than a month. Grafting is rarely practiced as the tree tends to sucker beneath the graft union. However, promising selections can be reproduced through leafy cuttings.

HARVEST AND USE The fruit ripens about a month after flowering. It is picked when fully colored. Storage characteristics are poor and the fruit begins to spoil within 2 or 3 days at room temperature. The cherry of the Rio Grande can be eaten out of hand or can be made into jams and preserves.

The tree is compact and beautiful and is amenable to many uses within the home landscape.

Chestnut

SCIENTIFIC NAME: *Castanea* spp.
FAMILY: Fagaceae
OTHER COMMON NAME: Castaña (Spanish)

Fruiting Calendar

JAN	FEB	MAR	APR	MAY	JUN	JUL	AUG	SEP	OCT	NOV	DEC

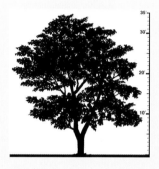

Characteristics

Overall Rating	★★★★
Ease of Care	★★
Taste/Quality	★★★★
Productivity	★★★★
Landscape Value	★★★
Wind Tolerance	★★★
Salt Tolerance	★
Drought Tolerance	★★★★
Flood Tolerance	★★★
Cold Tolerance	★★★★★

The chestnut will grow and fruit throughout north Florida and north central Florida. It makes an outstanding addition to the home garden and requires little care. The nut is a delicacy and is amenable to many uses. The burrs surrounding the nut make the area beneath the tree inaccessible to those with bare feet and can puncture mower tires. In locations where these concerns do not apply, the chestnut is an extremely rewarding species for dooryard planting.

Known Hazards

The nut case has sharp burrs.

GEOGRAPHIC DISTRIBUTION AND BACKGROUND

During early years of the last century, great forests of the American chestnut, *Castanea dentate,* carpeted the eastern United States. These huge trees each produced hundreds of pounds of small, sweet nuts. However, in 1904, chestnut blight was introduced from Asia. Within 40 years, the disease had wiped out all but a few isolated pockets of the American chestnut. Today, hundreds of thousands of bleached trunks still thrust skyward along Appalachian ridges, attesting to the durability of the wood. Little else remains. Various groups are working to back breed blight-resistant strains. The Chinese chestnut, *Castanea mollissima,* is resistant to chestnut blight. The nuts are medium to large and are of good quality. The European chestnut, *Castanea sativa,* is a large tree with medium-size nuts. It is susceptible to chestnut blight. The Japanese chestnut, *Castanea crenat,* is a medium-size tree that produces large nuts of poor quality.

TREE DESCRIPTION

The chestnut is a medium, deciduous tree. The American chestnut was a forest giant, sometimes exceeding 100 feet in height. Today's hybrids grow to 30 or 40 feet. Leaves are oblong-lanceolate, with deeply serrate margins. They measure from 5 to 10 inches in length. The tree will typically come into bearing about 5 or 6 years after it is planted. Male flowers are borne in catkins. Each female flower contains several carpels, which can develop into nuts. The chestnut is self-sterile. Two different cultivars should be planted to ensure pollination.

FRUIT CHARACTERISTICS

The nut is enclosed within a involucre or sharply spined burr, which splits at maturity. The shell is deep brown with a basal scar at the proximal end. Unlike most nuts, the chestnut is high in carbohydrates and low in fat. The cream-colored kernel is dense, sweet, and flavorful.

CULTIVARS

The Dunstan hybrid is a cross between blight-resistant American chestnuts and Chinese cultivars. It produces a nut of exceptional quality. 'Alachua,' 'Carolina,' 'Carpenter,' 'Heritage,' 'Revival,' and 'Willamette,' are named cultivars. 'Douglas' is another outstanding American-Chinese hybrid. Well-regarded Chinese cultivars include 'Abundance,' 'Carr,' 'Crane,' 'Kelsey,' 'Kuling,' 'Mendes,' 'Nanking,' and 'Qing.' Other hybrids that may hold promise for dooryard planting in north Florida are 'Colossal,' 'Eaton,' and 'Sleeping Giant.'

RELATIVES

The family Fagaceae encompasses many important species of forest trees. The beech, *Fagus grandifolia,* is a large tree native to north Florida. It bears small, brown, triangular nuts in a spiny husk. The oaks, *Quercus* spp., produce the familiar acorn, which is edible after tannins have been removed through an arduous process. The chinkapin, *Castanea* spp., is a dwarf chestnut native to Florida. It produces a small nut of good quality.

CLIMATE

The chestnut can withstand a wide range of climatological conditions, including severe cold. However, it requires at least 200 or 300 hours of chilling temperatures to set fruit. The tree is productive in Gainesville and Ocala; however, production becomes spotty further south. A tree in West Palm Beach bears half a dozen nuts annually.

CULTIVATION

Once established, the chestnut is an undemanding tree. It prefers well-drained, slightly acidic soil. A pH of about 6.0 is optimal. The chestnut is reasonably drought tolerant. Apart from chestnut blight, the tree is relatively free of pests and diseases. The chestnut gall wasp and chestnut weevil cause occasional problems. The most common method of propagation is grafting. This is carried out in the early spring, when rootstocks are dormant.

HARVEST AND USE

The average 10-year-old tree produces a crop of about 30 pounds. The usual method of harvest is to gather fallen nuts from around the base of the tree. Unless the nut is refrigerated or dried, it will deteriorate during extended storage. The chestnut can be eaten raw. It can also be roasted, grilled, steamed, sautéed, baked, boiled, or used as an ingredient in stuffing or stew. To prevent a buildup of pressure, the shell should be pierced prior to cooking.

Newly formed burrs and a catkin.

Immature burrs and dried catkins.

Chickasaw Plum

SCIENTIFIC NAME: *Prunus angustifolia*
FAMILY: Rosaceae

Fruiting Calendar

JAN	FEB	MAR	APR	MAY	JUN	JUL	AUG	SEP	OCT	NOV	DEC

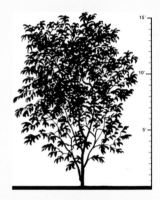

Characteristics

Overall Rating	★★
Ease of Care	★★★★★
Taste/Quality	★★
Productivity	★★★★
Landscape Value	★★★★
Wind Tolerance	★★★★
Salt Tolerance	★★
Drought Tolerance	★★★★
Flood Tolerance	★★★★
Cold Tolerance	★★★★★

The Chickasaw plum is a native fruit that was enjoyed by indigenous populations prior to the arrival of the Spaniards. The tree is ornamental, particularly when in bloom, and requires little maintenance. The fruit is of only marginal quality for eating out of hand, but is excellent in jellies and preserves. In addition to the Chickasaw plum, several varieties of low-chill commercial plums succeed in north and central areas of the state.

Known Hazards

The interior of the pit is toxic. Occasional sharp spines on terminal branches may cause injury.

GEOGRAPHIC DISTRIBUTION This species is native to the southeastern United States, growing as far west as Kansas and Oklahoma and as far north as New Jersey. In Florida, the Chickasaw plum occurs naturally as far south as Tampa. It succeeds as a dooryard tree from Orlando and Cape Canaveral northward. Authorities differ as to whether Native Americans introduced this species into Florida or whether its natural range included Florida. The tree occurs along fencerows, woodland borders, thickets, and disturbed areas. In north Florida it is sometimes used as a landscape planting for median strips and parking lots.

TREE DESCRIPTION The Chickasaw plum is a small tree, reaching about 20 feet in height. It may have multiple trunks or a single short trunk. Lateral branches are armed with thorns. The tree is deciduous, shedding its leaves from December until March. The alternate leaves are lanceolate and toothed, measuring from 1 1/2 to 3 inches in length. They are dark green above and pale green below. Immediately before the emergence of spring foliage, the tree is covered in clusters of fragrant, 1/2-inch, 5-petaled flowers. These are white with orange anthers. Growth is rapid, but the tree has a short life span.

FRUIT CHARACTERISTICS The fruit is a small, cherrylike drupe, measuring between 1/2 and 1 1/3 inches in diameter. The skin may be red or yellow. The juicy flesh has a hint of sweetness but is also somewhat sour. This tartness tends to dissipate when the fruit is fully ripe. The flesh contains a single stone.

CULTIVARS Efforts to improve the Chickasaw plum have been limited. The cultivar 'Guthrie' produces a large, yellow-skinned plum of good quality. 'Bruce' is a hybrid between the Chickasaw plum and the Japanese plum. The Chickasaw plum has been used for breeding purposes, to instill hardiness and disease resistance into several varieties of commercial plum.

RELATIVES The Chickasaw plum is a member of the Rosaceae family. Other members described within this book are the apple, *Malus domestica;* blackberry, *Rubus* spp.; capulin, *Prunus salicifolia;* loquat, *Eriobotrya japonica;* mayhaw, *Crataegus* spp.; peach, *Prunus persica;* pear, *Pyrus* spp.; and strawberry, *Fragaria ananassa.* The Chickasaw plum is probably the best of about a half dozen small-fruited native plums. These include the American plum, *Prunus americana;* Carolina laurelcherry, *Prunus caroliniana;* flatwoods plum, *Prunus umbellate;* and scrub plum, *Prunus geniculata,* an endangered species native to the central ridge area of central Florida. The European plum, *Prunus domestica,* and the Japanese plum, *Prunus salicina,* which actually originated in China, generally require significant winter chilling hours in order to set fruit. However, several cultivars will grow and fruit in north Florida. These include 'Black Ruby,' 'Blue Damson,' 'Burgundy,' 'Catalina,' 'Elephant Heart,' 'Methley,' 'Ruby Sweet,' and 'Segundo.' Low-chill varieties that will fruit in north central Florida include 'Gulfruby,' 'Gulfbeauty,' 'Gulfblaze,' and 'Gulfrose.'

CLIMATE The Chickasaw plum is not affected by the coldest temperatures likely to occur in Florida. It requires greater exposure to winter chill than is available in extreme south Florida.

CULTIVATION The Chickasaw plum has few cultural requirements. It will thrive on various soils, including infertile sand, so long as adequate drainage exists. However, it fairs poorly when planted in alkaline soil. The tree is drought tolerant. It should be pruned annually, after fruiting, to establish a compact yet accessible structure. No significant pests or diseases affect the tree. It is susceptible to vercillium wilt when planted on land formerly used for vegetable gardening. Tent caterpillars sometimes damage the foliage. The tree is generally grown from seed.

HARVEST AND USE The fruit typically ripens from late May through June. It can be eaten out of hand when fully ripe. It can also be used to make jellies, preserves, and pie fillings.

The plum flowers profusely in early spring. This is the 'Bruce' cultivar, popular in the Florida panhandle.

Coconut Palm

SCIENTIFIC NAME: *Cocos nucifera*
FAMILY: Arecaceae (Palmae)
OTHER COMMON NAME: Coco (Spanish)

Fruiting Calendar

JAN	FEB	MAR	APR	MAY	JUN	JUL	AUG	SEP	OCT	NOV	DEC

Characteristics

Overall Rating	★★★
Ease of Care	★★★★
Taste/Quality	★★★★
Productivity	★★★
Landscape Value	★★★★★
Wind Tolerance	★★★★★
Salt Tolerance	★★★★★
Drought Tolerance	★★★★
Flood Tolerance	★★★★
Cold Tolerance	★★

The coconut is a familiar tree in coastal south Florida. It is graceful, brings an exotic look to the landscape, and produces several crops annually. However, because the 'nut' is difficult to open, it is often disposed of as yard trash. Those who grow the tree should reacquaint themselves with the coconut as a food source. It is highly valued throughout most of the world's tropical regions.

Known Hazards

Falling fruit can cause blunt trauma injury and property damage.

GEOGRAPHIC DISTRIBUTION Some have speculated that the coconut originated in Malaysia. However, because the coconut is capable of transporting itself on ocean currents, its precise origin is uncertain. The coconut is widely distributed throughout the world's tropical and subtropical regions. Major centers of production include the Philippines, India, Indonesia, Sri Lanka, and Malaysia. Contrary to popular belief, the coconut palm does not need to grow near the ocean. It can tolerate salt but it does not require it. It will readily grow and fruit in lowland interior regions, if other climatic requirements are met. In the United States, the coconut is grown in Florida and Hawaii.

TREE DESCRIPTION The coconut may achieve a height of between 30 and 100 feet, depending on the variety. The trunk, often curved and swollen at the base, has smooth gray bark, marked with indistinct leaf scars. The leaves or fronds are pinnate and may measure up to 16 feet in length. The coconut palm is an extremely fast-growing tree. If given proper care, a container-grown specimen may reach 12 feet in height within 3 years. The tree typically begins bearing at between 3 and 7 years of age.

The tree flowers several times during the year, more frequently over the warmer months. The tough-stemmed inflorescence is about 3 feet in length. It typically bears several thousand flowers. Most are male. Female flowers are large, and form close to the central stem of the inflorescence. Each inflorescence is capable of producing up to 2 dozen nuts.

The coconut palm can support tremendous crops. The fruit matures throughout the year.

FRUIT CHARACTERISTICS The fruit of the coconut is not a true nut. It is a large, dry, fibrous drupe. The husk is composed of the thin exocarp and thick mesocarp. It is reinforced with tough, longitudinal fibers. At the center is a brittle-shelled endocarp, often thought of as the nut. This contains endosperm tissue (the "meat"), liquid endosperm (the "water"), and a tiny plant embryo. The fruit can measure up to 18 inches in diameter and can weigh more than 35 lbs. If the nut is harvested 7 to 8 months after flowering, it will contain liquid (coconut water) surrounded by jelly-like pulp. If the nut is harvested 11 or 12 months after flowering, it will contain a small amount of liquid surrounded by thick layer of white "meat."

CULTIVARS The species falls into 2 geographic groups: coconuts from West Africa and India, and coconuts from the Pacific region and Southeast Asia. At least 100 strains or cultivars exist. In Florida, cultivars must be selected based on their resistance to the fatal disease, lethal yellowing. 'Jamaica Tall' was once commonly planted in south Florida but has been nearly wiped out by the disease. Cultivars that have shown some resistance, and that are sometimes available in Florida, include 'Fiji Dwarf,' 'Maypan,' 'Malayan Green Dwarf,' 'Malayan Yellow Dwarf,' 'Malayan Golden Dwarf,' and 'Spicata Dwarf.'

RELATIVES The Arecaceae or Palm family, within the order Arecales and the subclass Arecidae, contains approximately 2,500 species. It comprises one of the most important plant families in terms of human use. The list of edible palms is extensive. Several are well adapted to Florida.

The jelly or pindo palm, *Butea capitata,* native to South America, is a familiar landscape plant throughout central and north Florida. While often praised for its beauty, it is seldom recognized for its potential as a fruit tree. The fruit is wholesome and flavorful. Bunches may exceed 30 pounds in

The pindo palm is hardy in north Florida.

weight. The jelly palm has survived temperature drops to 10° F. For those who seek to grow a fruiting palm in areas too cold for the coconut, the jelly palm is an excellent choice.

The peach palm or pejibaye, *Bactris gasipaes,* is native to South America. It may grow singly, but more often forms a cluster, suckering from the base. It will grow in south Florida, but is frost sensitive. The trunk is covered with sharp, downward-pointing spines that complicate harvest. The reddish-orange fruit is born in large clusters. It is boiled in saltwater before it is consumed.

This extensive collection of peach palms is located at CATIE (the Tropical Agriculture Research and Higher Learning Center) in Costa Rica.

The African oil palm is tough and productive.

The African oil palm, *Elaeis guineensis,* is a palm native to Central Africa that attains a height of about 40 feet. It bears prolific quantities of fruit, which are renowned for their oil-rich pulp. The fruit are egg-shaped, about 2 inches in diameter, resembling small, golden-orange coconuts. This palm is reasonably well adapted to extreme South Florida and can tolerate poorly drained soils.

The queen palm, *Syagrus romanzoffiana,* is native to South America. It is, perhaps, the most widely planted landscape palm in south and central Florida. The fruit is fibrous and the flesh is thin. Yet it is also flavorful, resembling the canistel in taste and consistency.

The date palm, *Phoenix dactylifera,* native to North Africa, is frequently used for landscape purposes in Florida. However, it is rarely, if ever, cultivated for its fruit. High humidity interferes with fruit set and often causes the fruit to drop prematurely. Nevertheless, where a male pollinator is present, the date palm will occasionally produce a worthwhile crop.

These date palms, in Jupiter, Florida, set a heavy crop. However, no male pollinator was present.

CLIMATE

The coconut requires a tropical or near tropical climate. In Florida, it is a good indicator tree. Where the coconut grows to maturity, many other tropical species will survive. Foliage will suffer damage from a light frost. This shows up as a general browning and necrosis of outer fronds. A temperature drop to 27° F will kill the central bud. On the east coast, the coconut grows on barrier islands as far north as Cocoa Beach. However, most specimens north of Stuart were planted after the disastrous freeze of 1989. On the Gulf coast, the coconut is occasionally planted as far north as Sarasota.

CULTIVATION

The coconut palm is an extremely low-maintenance tree. It prefers a well-drained, sandy loam but will tolerate extremely poor soil. It has remarkable salt tolerance and will grow within 3 feet of mean high tide. The coconut palm can also endure significant periods of drought. However, a young tree should receive regular irrigation for several months until it is established.

The tree benefits from twice-annual fertilization with an 8–3–10 fertilizer or some similar formula. Potassium deficiency is fairly common and manifests itself as a flecking or spotting and necrosis on older leaves. It can eventually lead to a constriction of the trunk or to the death of the tree. Yellow bands along the margins of older fronds are an indication of magnesium deficiency.

The constriction or "penciling" in the trunk of this 'Malaysian Dwarf' coconut resulted from a potassium deficiency.

Site selection is important. Coconut palms should never be planted over patios, driveways, roads, pools, benches, or areas where people congregate. The nut can fall without warning on a breezeless day. Falling coconuts can dent vehicles and shatter windshields. They have been known to seriously injure people. It is important to remove the nuts when a hurricane or tropical storm threatens. Hundred-mile-per-hour winds can turn the nut into a cannonball—easily capable of punching out windows and penetrating siding.

PESTS AND DISEASES Lethal yellowing is by far the most serious disease of the coconut in south Florida. It was discovered in the Caribbean in the 1800s. Scientists believe that the disease is caused by a phytoplasma, an organism not much larger than a virus. They suspect that it is spread by a planthopper. At least 30 other species of palm are susceptible to the disease. Early symptoms include premature fruit drop and browning or darkening of the inflorescence. Soon, the fronds begin to yellow. Finally, the center bud dies. The best preventive measure is to select varieties resistant to the disease. Unfortunately, many cultivars that were once thought to be immune have proven to be susceptible.

Bud rot is a fungal disease common in areas of high humidity and rainfall. It is generally lethal. The leaves turn brown and wither as the bud decays. Pests include scales, mealybugs, aphids, and palm weevils. Squirrels and rats sometimes gnaw holes in the husk of the fruit to get at the bounty within.

PROPAGATION The coconut palm is generally grown from seed. Although the tree can be planted at any time, it will establish most readily if it is planted in May, at the start of the rainy season.

HARVEST AND USE In Florida, the coconut matures throughout the year. The nuts are severed from the tree with a pole saw. Opening the fruit can be a daunting challenge. In fact, without some guidance and experience it can be downright dangerous.

The most difficult task is to remove the husk. Commercial de-husking machines are expensive. One method of removing the husk relies on a metal stake driven into the ground or other surface. A sharp spike is left protruding 3 inches into the air. The coconut, held firmly in both hands, is driven downward and impaled on the point. It is than twisted back and forth to separate the fibers of the husk.

Several methods can be used to open the nut once it has been removed from the husk. One technique is to hold the nut over a bowl and strike it repeatedly with the blunt edge of a heavy cleaver. If the nut is rotated slowly, and if the blows are registered in a line around the middle, the nut will eventually crack in half. The liquid can then drain into the bowl below. The shell can also be pierced to retrieve the coconut water. A screwdriver can be tapped into one of the 'eyes' and the liquid drained from within. However, the opening may become plugged with pieces of endocarp. The result is often a maddeningly slow trickle of liquid. Once the liquid has been removed, the coconut can be heated for an hour in an oven, then wrapped in a towel and hit with a hammer. This will cause the shell to shatter. A knife is then used to free the meat from the shell.

The coconut lends itself to an incredible range of uses. Coconut water makes a delicious and refreshing drink. Coconut milk is an emulsion of the pressed or pulverized endocarp and water. It is widely used in cooking. Combining the endosperm and coconut water in a blender makes an acceptable product. The mature endosperm or meat can be grated and used in various recipes.

Cocoplum

SCIENTIFIC NAME: *Chrysobalanus icaco*
FAMILY: Chrysobalanaceae
OTHER COMMON NAME: Hicaco (Spanish)

Fruiting Calendar

JAN	FEB	MAR	APR	MAY	JUN	JUL	AUG	SEP	OCT	NOV	DEC

Characteristics

Overall Rating	★★
Ease of Care	★★★★★
Taste/Quality	★★
Productivity	★★
Landscape Value	★★★★★
Wind Tolerance	★★★★
Salt Tolerance	★★★★
Drought Tolerance	★★★★
Flood Tolerance	★★★
Cold Tolerance	★★

The cocoplum is a native species that is widely planted as an ornamental hedge in south Florida. While the fruit is of only fair quality, it can be made into a unique jelly. The real treasure is the rich seed endocarp. The plant is not cold tolerant. However, it is drought tolerant and resilient. It is useful in seaside landscaping due to its salt tolerance. The cocoplum's versatility makes it a valuable addition to any south Florida garden.

Known Hazards

None

GEOGRAPHIC DISTRIBUTION The cocoplum is native to south Florida, the West Indies, and the Caribbean coast of Mexico, Central America, and South America. In Florida, it occurs naturally in cypress hammocks, wetland areas, disturbed areas, roadsides, and costal dunes. It will grow very close to the beach but is equally at home further inland. The cocoplum was widely gathered for food by Native Americans prior to the arrival of the Spaniards in Florida.

PLANT DESCRIPTION The cocoplum is an evergreen shrub. Without training, it forms a dense, rounded, clumping bush. It can attain a height of about 10 feet and a spread of 20 feet or more. It can be trained into an attractive hedge, which is its primary landscape use in Florida. The cocoplum is low branching. Canelike new branches often emerge directly from the roots. Leaves are dark green, leathery, shiny, rounded to obovate. The cocoplum produces tiny, fragrant, white flowers over much of the year.

FRUIT CHARACTERISTICS The fruit of the cocoplum is a small, ovoid drupe, measuring about 1 1/2 inches in diameter. The skin is thin and tender and encloses a layer of juicy, white, somewhat cottony pulp. The flavor is sweet and pleasant, but somewhat insipid. A single seed, approximately the size and shape of a pistachio nut, is embedded in the flesh. Inside the brittle shell is a cream-colored kernel. The kernel has a unique and pleasant flavor, with hints of almond and coconut.

VARIETIES Several strains are available. The Hobe Sound cocoplum is a diminutive plant, often used as a low hedge. The red-tip cocoplum forms a fairly large plant. Emerging foliage is scarlet in color. The Everglades cocoplum is said to favor wet areas. All produce fruit of similar quality.

RELATIVES The cocoplum belongs to the small Chrysobalanceae family. Another member, the gopher apple, *Licania michauxii,* is a low-growing native plant that bears a fruit of minor import. The plant has greater cold tolerance than the cocoplum and is sometimes used as a groundcover. Several Chrysobalanceae species are native to Central America. The sunsapote, *Licania platypus,* grows in Mexico and Central America. It produces a large, brown-skinned fruit of mediocre quality.

CLIMATE The foliage of the cocoplum will freeze at about 27° F. So long as temperatures do not fall below 23° F, the plant will usually resprout. The cocoplum occurs as far north as Brevard County on the east coast and Sarasota on the Gulf coast.

CULTIVATION Once established, the cocoplum requires little care. It is drought tolerant, but may require irrigation during establishment. It does not require fertilization even when grown in very poor soil. It establishes readily on windy sites. It has moderate resistance to salt. The cocoplum is usually grown from seed. Seeds require between 5 and 10 weeks to germinate. It can also be propagated through cuttings and air layering. Few pests or diseases affect the cocoplum.

HARVEST AND USE Fruit production peaks from May through August, although some plants fruit sporadically throughout the year. The fruit can be consumed out of hand, but is of only fair eating quality. The seed kernel or "nut" is highly prized. It can be eaten raw or it can be slow roasted, preferably with salt and a dab of coconut oil. Cocoplum jelly is prepared by placing skinned plums in water just deep enough to cover the top layer of fruit. A generous quantity of sugar is added and the mixture is brought to a boil, then left to simmer for several hours. The stew is then roughly strained. The seeds are cracked and the endocarps are crushed and sprinkled into the mixture, which is reduced further. The finished jelly is then permitted to cool and congeal.

The tiny flowers of the cocoplum in clusters on the branch tips.

Custard Apple

SCIENTIFIC NAME: *Annona reticulata*
FAMILY: Annonaceae
OTHER COMMON NAMES: Bullock's Heart,
    Corazón (Spanish)

Fruiting Calendar

JAN	FEB	MAR	APR	MAY	JUN	JUL	AUG	SEP	OCT	NOV	DEC

Characteristics

Overall Rating	★★
Ease of Care	★★★
Taste/Quality	★★
Productivity	★★★
Landscape Value	★★
Wind Tolerance	★★★
Salt Tolerance	★★
Drought Tolerance	★★★
Flood Tolerance	★★★
Cold Tolerance	★★

The fruit of the custard apple is regarded as mediocre when compared with that borne by more glamorous members of the Annona family. Nevertheless, the species has several advantages. It bears fruit over the late winter and early spring when few other crops are available. It is usually productive without hand pollination. The average fruit is large and can be of good quality. For these reasons, the custard apple deserves to be planted in south Florida where space will allow.

Known Hazards

The seed contains toxins. The sap may be an irritant.

GEOGRAPHIC DISTRIBUTION The custard apple may have originated in the Antilles, although related species grow in Guatemala and Belize. It was widely distributed throughout Central America prior to the arrival of the Europeans. It is grown in Mexico, South Africa, India, Sri Lanka, Malaysia, and the Philippines.

TREE DESCRIPTION The custard apple is a deciduous tree of small to medium stature. It may attain a height of between 15 and 30 feet. The foliage is course in texture and the crown is irregular in form. Leaves are alternate, oblong-lanceolate, measuring 4 to 7 inches in length. In Florida, the tree sheds most of its leaves over the cooler months. The trunk is covered in smooth, gray bark. The custard apple makes a fairly attractive landscape specimen. Fallen fruit can be a nuisance and can draw yellow jackets and other insects. The flower is similar to that of the atemoya.

FRUIT CHARACTERISTICS Like many fruit-bearing members of the Annona family, the custard apple produces a compound fruit. The fruit is ovate, somewhat heart-shaped, and usually measures from 4 to 7 inches in diameter. The thick stem is inserted into a depression at the base, and continues as an inedible core through the center of the fruit. The skin varies in color and may be red, purple, brown, bronze, or yellow. It is vaguely marked by hexagonal areoles.

The flesh is usually off-white and is sometimes tinged with pink near the rind. Immediately beneath the rind is a layer of somewhat granular flesh. Toward the core, the flesh is divided into poorly defined carpels, which range in consistency from soft and juicy to gelatinous and stringy. Most segments enclose a single hard seed. In superior selections, the flavor is sweet and aromatic. In inferior varieties, the flavor is insipid and the texture is grainy.

CULTIVARS Several superior cultivars have been selected. These include 'Benque,' which produces a large fruit with dark-pink pulp; 'Canul,' which produces a medium-size fruit with red skin and purplish red pulp; 'Chonox,' which produces a medium-size fruit with pink flesh; 'San Pablo,' which produces a large, high-quality fruit with pink flesh; 'Sartenaya,' a well regarded variety which produces a medium-size fruit with pink flesh; and 'Tikal,' a variety said to produce a fruit of very good quality.

RELATIVES Other Annonaceae species discussed within this book include the atemoya, *Annona cherimola* x *Annona squamosa;* the pawpaw, *Asimina triloba;* and the sugar apple, *Annona squamosa.*

The subsection pertaining to relatives of the atemoya describes other related species.

CLIMATE The custard apple requires a tropical or near tropical climate. In Florida, it grows as far north as Ft. Pierce on the east coast and Sarasota on the west coast. Foliage is damaged at about 29° or 30° F. Mature specimens are capable of withstanding temperature drops to about 27° F.

CULTIVATION The custard apple is a hardy tree requiring little care. It is not particular as to soil type, growing in sand and on the oolitic limestone of Miami-Dade County. It does, however, require proper drainage. The custard apple requires irrigation during periods of drought. The fruit is subject to attack by the chalcid wasp. Grafting results in a high rate of success and the side-veneer graft is frequently employed.

HARVEST AND USE The custard apple comes into bearing in late winter and spring. A mature, well-tended tree may bear 100 pounds of fruit. The fruit is harvested when it has attained full size and lost its green color. The fruit softens slightly as it ripens, with the skin yielding to gentle pressure. The fruit is often eaten fresh, with the flesh spooned from the rind and the seeds discarded. The flesh may also be used as an additive to ice cream or milkshakes—although care must be taken to exclude the seeds, which are poisonous.

The tree of the custard apple has a sprawling habit of growth, similar to that of many Annonas.

Custard Apple

Darling Plum

SCIENTIFIC NAME: *Reynosia septentriolalis*
FAMILY: Rhamnaceae
OTHER COMMON NAME: Red Ironwood

Fruiting Calendar

JAN	FEB	MAR	APR	MAY	JUN	JUL	AUG	SEP	OCT	NOV	DEC

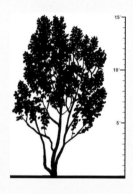

Characteristics

Overall Rating	★★★
Ease of Care	★★★★
Taste/Quality	★★
Productivity	★★
Landscape Value	★★★★
Wind Tolerance	★★★★
Salt Tolerance	★★★
Drought Tolerance	★★★★
Flood Tolerance	★★★
Cold Tolerance	★★

The Darling plum is a beautiful native shrub that bears a fruit of good quality. The plant is tough and resilient, capable of growing under adverse conditions. As a result of its moderate salt tolerance, the Darling plum can be grown on barrier islands. The fruit is thin fleshed but of very fine flavor. Residents of south Florida should consider incorporating the Darling plum into the home landscape as a suburban multi-purpose plant.

Known Hazards

None

GEOGRAPHIC DISTRIBUTION The Darling plum is native to south Florida and the Caribbean. It is found in coastal hammocks as far north as Cape Canaveral along Florida's east coast. Within its native habitat, the Darling plum is considered a threatened species and is being regularly displaced by coastal development.

TREE DESCRIPTION The Darling plum is a medium, sprawling, evergreen shrub, capable of attaining a height of 15 to 20 feet. The deep green leaves are leathery and entire. They are obovate and slightly emarginate. New foliage has an attractive reddish tint. The wood is dark, hard, and durable. The plant has a slow growth rate. Flowers are inconspicuous, small, and greenish-yellow, with 5 prominent sepals and no petals.

FRUIT CHARACTERISTICS The fruit is a small drupe, green when immature, turning red, then purple-black as it ripens. In outward appearance the fruit resembles that of the cocoplum, but has a spine at the apex. It ranges from about 3/4 of an inch to 1 inch in length. The skin is of medium thickness and contains a thin layer of juicy flesh. The taste is sweet, resembling that of a pear or blueberry. Each fruit contains a single yellowish stone, which constitutes more than 50 percent of the overall weight.

CULTIVARS To date, no selections of Darling Plum have been made, and propagation is primarily by seed. Potted specimens are available from a scattering of nurseries that specialize in native plants. Efforts should be undertaken to select specimens with larger fruit, smaller stones, and superior fruiting characteristics.

RELATIVES The Rhamnaceae or Buckthorn family contains about 600 species in about 50 genera. The jujube, *Ziziphus* spp., addressed within a separate section of this book, probably represents the most important fruiting member of the family. The black ironwood, *Krugiodendron ferreum,* is a native Rhamnaceae that bears edible fruit. This hardy and well-behaved tree bears an abundant crop of small, sweet drupes over the course of several months. It is similar to the Darling plum in its habit of growth. The black ironwood makes an excellent small specimen tree and deserves wider planting. Clusters of insignificant, yellow flowers appear in late spring and early summer. Fruiting peaks in September and November, but sporadic fruiting may occur from early summer through early winter. Other native species include the Carolina buckthorn, *Rhamnus caroliniana;* the endangered Cuban nakedwood, *Colubrina cubensis;* littleleaf buckbrush, *Ceanothus microphyllus;* the endangered scrub jujube, *Ziziphus celata;* and the endangered soldierwood, *Colubrina elliptica.*

CLIMATE The Darling plum is tropical or subtropical in habit. While is uncertain how much cold the species can withstand, it would probably be unwise to attempt cultivation far from coastal areas of south Florida.

CULTIVATION The Darling plum is an extremely tough and rugged plant. It is not particular as to soil type and will withstand a wide pH range. It is drought tolerant and is somewhat salt tolerant. The plant rarely suffers from any nutritional deficiencies. Water should be provided for several months following planting. Once established, the Darling plum requires little or no care. The species prefers full sun, but will grow and fruit in partial shade. Few pests or diseases affect the species. The Darling plum is grown from seed.

HARVEST AND USE Fruit production, which occurs sporadically over the late summer and fall, is moderate. The fruit is picked when it achieves full coloration and is usually eaten out of hand. It can also be used in preserves.

Elderberry

SCIENTIFIC NAME: *Sambucus* spp.
FAMILY: Caprifoliaceae or Adoxaceae
OTHER COMMON NAME: Baya del Saúco (Spanish)

Fruiting Calendar

JAN	FEB	MAR	APR	MAY	JUN	JUL	AUG	SEP	OCT	NOV	DEC

Characteristics

Overall Rating	★★
Ease of Care	★★★★
Taste/Quality	★
Productivity	★★★
Landscape Value	★★★
Wind Tolerance	★★★★
Salt Tolerance	★
Drought Tolerance	★★★
Flood Tolerance	★★★★
Cold Tolerance	★★★★★

The elderberry is a common native fruit of minor import. The fruit is small and is not suited to fresh consumption. However, when cooked, it makes well-regarded preserves and pie fillings. The plant is hardy, disease resistant and easy to grow. The white flower, present throughout much of the year, forms showy clusters. For those who favor native plants, the elderberry makes an interesting and productive addition to the landscape.

Known Hazards

All parts of plant are poisonous; uncooked fruit are poisonous.

GEOGRAPHIC DISTRIBUTION The common elderberry is widely distributed throughout the eastern United States, ranging as far north as Nova Scotia and as far south as Miami. It frequents ditch banks, roadsides, fence lines, and low hollows.

TREE DESCRIPTION The elderberry is a deciduous shrub, attaining a height and spread of from 6 to 14 feet. It is somewhat rangy or willowy in habit, often forming multiple trunks. Leaves are pinnately compound, typically composed of between 5 and 11 leaflets. The leaflets measure from 2 to 3 inches in length and have serrated margins. The elderberry often forms dense stands and may reproduce by sending out runners. Roots are fibrous and shallow. Stems are hollow, filled with white pith.

Within the home garden, the elderberry is usually planted as part of a wildlife garden, naturalized area, or roadside screen. The plant is attractive when in bloom. The small, white flowers are typically composed of a 5-lobed corolla and 5 stamens. They measure about 1/4 of an inch across. The flowers form dense, flat-topped clusters measuring from 6 to 9 inches in diameter. The plant typically blooms in late spring or early summer, although in south Florida, flowering may occur at any time.

FRUIT CHARACTERISTICS Flowering and fruiting take place primarily on 2-year old canes. The fruit is a small, globose, berrylike drupe, ranging in color from red to purple to black. It rarely measures more than 1/3 of in inch in diameter. The pulp is juicy and contains from 1 to 5 small nutlets. The flavor is sweet, but the raw fruit is insipid, unpalatable, and has been known to cause digestive upset.

SPECIES AND CULTIVARS The elderberry encompasses several species including the common elderberry, *Sambucus Canadensis;* the blue elderberry, *Sambucus caerulea;* and the black elderberry, *Sambucus nigra.* Where selected clones are grown, 2 or more varieties should be planted in close proximity to ensure pollination. Cultivars for grown their fruit include 'Adams,' 'Aurea,' 'Johns,' 'Kent,' 'Nova,' and 'York.' Because most of these cultivars were selected in northern states, performance under Florida conditions is uncertain.

RELATIVES The elderberry belongs to the Caprifoliaceae or Honeysuckle family. Besides the elderberry, several species native to Florida belong to this family, including the coralberry, *Symphoricarpos orbiculatus;* coral honeysuckle, *Lonicera sempervirens;* possumhaw, *Viburnum nudum;* rusty blackhaw, *Viburnum rufidulum;* and small-leaf viburnum, *Viburnum obovatum.* Some authorities have divided the Caprifoliaceae family and have placed the elderberry within the associated Adoxaceae family.

CLIMATE The elderberry is not affected by the coldest temperatures likely to occur in Florida.

CULTIVATION The elderberry is easy to maintain and will tolerate a wide range of growing conditions. It prefers rich, moist soil, but also performs well in sandy soil. It prospers in full sun but can endure partial shade. Branches over 3 years old should be removed, as these rarely flower or fruit. Fertilization is not required. Heavy applications of nitrogen lead to excessive vegetative growth at the expense of fruit production. The elderberry has few serious pests or diseases. It is frequently propagated from seed, which is sown shortly after removal from the ripe fruit. Cuttings from the current season's growth have been successfully rooted. In addition, suckers may be separated from the parent plant.

HARVEST AND USE Peak fruit production occurs in late summer or fall. The elderberry contains hydrocyanic acid. All parts of the plant, including the raw fruit, are poisonous if ingested. Cooking destroys and eliminates this toxin. The fruit has various culinary uses and may be substituted for blueberries in muffins and other baked goods. The fruit can be used to make jellies, sauces, pie filling, and wine.

The elderberry is a frequent inhabitant of roadside ditches.

Feijoa

SCIENTIFIC NAME: *Feijoa sellowiana*
FAMILY: Myrtaceae
OTHER COMMON NAME: Pineapple Guava,
　　Guayabo del Pais (Spanish)

Fruiting Calendar

JAN	FEB	MAR	APR	MAY	JUN	JUL	AUG	SEP	OCT	NOV	DEC

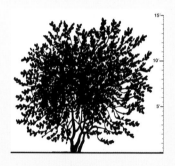

Characteristics

Overall Rating	★★★
Ease of Care	★★★
Taste/Quality	★★★★
Productivity	★★★
Landscape Value	★★★★★
Wind Tolerance	★★★
Salt Tolerance	★★★
Drought Tolerance	★★★
Flood Tolerance	★★★
Cold Tolerance	★★★★★

The feijoa produces a delicious fruit and ornamental flowers with edible petals. It is an outstanding plant for the home landscape. Fruit quality declines marginally with exposure to high temperatures, although the fruit produced by this species in Florida is of very good quality. The feijoa is beautiful and productive and merits planting as a dooryard crop in north and central Florida.

Known Hazards

None

GEOGRAPHIC DISTRIBUTION The feijoa is native to upland areas of Uruguay, southern Brazil, and northern Argentina. It is a commercial crop in California, Uruguay, Israel, Republic of Georgia, Australia, and New Zealand.

PLANT DESCRIPTION The feijoa is an evergreen shrub, attaining a height of about 15 feet. It may be upright or somewhat sprawling in form. The foliage is ornamental. The leathery leaves are opposite, deep green above, silvery-white below. They measure between 1 and 2 1/2 inches in length and are elliptical in form. Flowers are showy, averaging about an inch in diameter. The fleshy petals, usually 4 in number, are reddish-pink with white undersides. A spray of bright red stamens emanates from the center of the flower. In Florida, bees appear to be the chief pollinating agents. However, in New Zealand and other locations, birds eat the flower petals and transfer the pollen on their heads. Many cultivars require cross-pollination from a different cultivar to set fruit.

FRUIT CHARACTERISTICS The fruit is ovate or pyriform, measuring between 1 and 3 1/2 inches in length. A persistent calyx is present at the apex. The skin is dull, blue-green, and inedible. Within the rind lies the cream-colored flesh, somewhat pearlike in texture, enclosing several dozen small seeds that do not interfere with eating. The center becomes translucent and gelatinous as ripening progresses. The flavor is sweet, aromatic, musky, and pleasant, and is thought by some to resemble pineapple—although the resemblance is not striking.

CULTIVARS Cultivars that are sometimes available in Florida include, 'Nazemetz,' a self-fertile California cultivar that bears large, sweet fruit of excellent flavor; 'Apollo,' a self-fertile New Zealand cultivar that bears large fruit of very good quality; 'Gemini,' a New Zealand cultivar that produces small fruit of high quality; 'Improved Coolidge,' a California cultivar that bears large fruit of good quality; and 'Coolidge' a self-fertile Australian cultivar that produces heavy crops of fair quality fruit. New Zealand cultivars that would be worth attempting in Florida include 'Kakapo,' 'Mammoth,' 'Monique,' 'Opal Star,' 'Pounamu,' 'Triumph,' 'Unique,' and 'White Goose.' California cultivars that would be worth attempting in Florida include 'Beechwood,' 'Lickver's Pride' 'Pineapple Gem,' and 'Smilax.'

RELATIVES Several members of the family Myrtaceae are discussed in this book, including the blue grape, *Myrciaria vexator;* cattley guava, *Psidium cattleianum;* cherry of the Rio Grande, *Eugenia aggregata;* grumicahma, *Eugenia braziliensis;* guava, *Psidium guajava;* jaboticaba, *Myrciaria* spp.; pitomba, *Eugenia luschnanthiana;* and stoppers, *Eugenia* spp.

CLIMATE The feijoa does not fruit well in extreme south Florida, as it requires about 100 chilling hours to set an adequate crop. It is capable of surviving a temperature drop to about 12° F.

CULTIVATION The feijoa is a hardy, low-maintenance plant. It will grow in various soils, but requires good drainage. It is drought tolerant. However, irrigation should be provided during establishment and during fruit development. Because extreme heat has a negative impact on fruit quality, the feijoa might benefit from midday shade in Florida. Frequent, light applications of a balanced fertilizer are recommended. The feijoa is an occasional host of the Caribbean fruit fly, although the rate of infestation appears to be low. The preferred method of propagation is to root softwood cuttings. Grafting is impracticable because the plant suckers beneath the graft union.

HARVEST AND USE In Florida, fruit ripen from mid-summer through October. Under ideal conditions, a mature specimen will produce about 1,000 fruit per year. Ripe fruit can be shaken from the tree or gathered from below after it has already fallen. The fruit is ready to eat when it is slightly soft to the touch. Once cut, the fruit should be quickly consumed as the flesh discolors with exposure to air. Flower petals are also edible. They taste faintly of cinnamon. As long as central parts of the flower are not disturbed, harvesting the petals does not significantly reduce fruit set.

The foliage of the feijoa is attractive and the plant has tremendous landscape appeal.

Fig

SCIENTIFIC NAME: *Ficus carica*
FAMILY: Moraceae
OTHER COMMON NAME: Higo (Spanish)

Fruiting Calendar

JAN	FEB	MAR	APR	MAY	JUN	JUL	AUG	SEP	OCT	NOV	DEC

Characteristics

Overall Rating	★★★★
Ease of Care	★★
Taste/Quality	★★★
Productivity	★★★
Landscape Value	★★★
Wind Tolerance	★★★
Salt Tolerance	★★
Drought Tolerance	★★★
Flood Tolerance	★★★
Cold Tolerance	★★★★★

The fig is one of the most popular and desirable fruits of warm-temperate areas. It succeeds in every region of Florida, from Pensacola to Homestead. Once it is established, the fig requires little maintenance, although it may be troubled by nematodes when planted in sandy soil. The tree is handsome and well behaved. Some cultivars produce 2 crops a year. In light of its many attributes, the fig should be used with great frequency in the home landscape.

Known Hazards

The latex from fruit stems can cause skin irritation and dermatitis.

92

GEOGRAPHIC DISTRIBUTION The fig is native to Asia Minor or the Middle East. It has been cultivated for thousands of years and is prominent in the history and culture of Greece and other Mediterranean nations. The fig was grown as a commercial crop in Florida during the late 1800s.

TREE DESCRIPTION The fig is a deciduous small tree or spreading bush. It rarely attains a height of greater than 25 feet. Leaves are variable. They are broad and palmate reaching up to 10 inches across. Most are roughly toothed and deeply notched forming 3 or 5 lobes. The trunk, covered with thin gray bark, rarely exceeds 8 inches in diameter. The root system is greedy and can reach out over a considerable distance. The fig makes a handsome landscape specimen.

The shape of the fig leaf varies with the cultivar.

FRUIT CHARACTERISTICS The fig is not a true fruit, but is a synconium, a hollow stemlike structure, bearing flowers along its interior walls. A small osteole, commonly referred to as the eye, is located at the apex. The interior is lined with thousands of tiny flowers that impart the characteristic sweetness and flavor. The entire fig—flowers, fleshy wall, and skin—is edible.

CULTIVARS The species *Ficus carica* is composed of 4 types of figs: the common fig, the caprifig, the Smyrna fig, and the San Pedro fig. Only the common fig is of import to growers in Florida. At least 120 cultivars exist. Among the best for dooryard cultivation in Florida are 'Green Ischia,' 'Kadota,' 'Celeste,' 'Brown Turkey,' 'Black Mission,' 'Excell,' 'Alma,' 'Nero,' 'Desert King,' 'Brunswick,' 'Conadria,' 'Flanders,' 'Hunt,' 'Kalamata White,' 'LSU Gold,' 'LSU Purple,' 'Ventura,' and 'White Marseille.'

RELATIVES The fig belongs to the diverse Moraceae family. Moraceae species covered within this book are the che, *Cudrania tricuspidata;* mulberry, *Morus* spp.; and jackfruit, *Artocarpus heterophyllus.* Many members of the *Ficus* genus are huge rubber trees native to tropical rainforests. Few bear fruit that compares qualitatively with that of the common fig. Notable fruiting species include the cluster fig, *Ficus racemosa,* of Southeast Asia; the sycamore fig, *Ficus sycomorus,* of Africa and the Middle East; and the sandpaper fig, *Ficus coronata,* of Australia.

CLIMATE The fig is hardy throughout Florida. It can withstand temperatures as low as 10° F when dormant. However, a light frost can damage a tree that is in an active state of growth. Some varieties have moderate chilling requirements, but most are productive in south Florida.

CULTIVATION The fig is not particular as to soil, but requires good drainage. It will withstand minor salinity in the soil and water. The plant is drought tolerant, but is more likely to produce good crops with periodic irrigation. The fig should not be planted near a septic tank drain field, as its fibrous roots can clog lines. When young, the plant should be protected from direct sun with shade cloth. The tree may benefit from occasional light applications of 5–10–10 fertilizer. Excessive nitrogen reduces fruit production. The plant is pruned back once the main crop has been harvested. The fig is susceptible to root damage from nematodes. This damage can be reduced through heavy mulching or by planting the fig near a structure where the roots can shelter beneath the footer. The species can be propagated through hardwood cuttings, air layering, and ground layering.

HARVEST AND USE A mature, productive tree may bear 100 pounds of fruit. Some varieties bear a light crop in the spring called the breba crop. The main crop is borne in late summer or early fall. The fruit is deemed mature when it droops slightly and separates easily from the branch. Sticky latex that oozes from the plant is an irritant, so gloves should be worn during harvest. The fruit can be eaten out of hand, dried, baked, frozen, or used in preserves.

Compare four fig cultivars, from left to right: 'Kadota,' 'Brown Turkey,' 'Green Ischia,' and 'Black Mission.'

Fig

93

Grapefruit

SCIENTIFIC NAME: *Citrus* x *paradisi*
FAMILY: Rutaceae
OTHER COMMON NAME: Toronja (Spanish)

Fruiting Calendar

JAN	FEB	MAR	APR	MAY	JUN	JUL	AUG	SEP	OCT	NOV	DEC

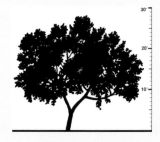

Characteristics

Overall Rating	★★★★
Ease of Care	★★★★
Taste/Quality	★★★★
Productivity	★★★★
Landscape Value	★★★
Wind Tolerance	★★★
Salt Tolerance	★★
Drought Tolerance	★★★★
Flood Tolerance	★★★
Cold Tolerance	★★

The grapefruit is a familiar Citrus species with many attributes and few drawbacks. It is well adapted to Florida growing conditions and requires little maintenance. It provides an extended harvest season. The fruit lends itself to many uses. The grapefruit may not have the exotic appeal of many of the species discussed within this book, but it is a steady workhorse, capable of producing a delicious and dependable crop over many years.

Known Hazards

The rind and leaves contain oils that have been known to cause dermatitis in some individuals. Most varieties have occasional thorns.

GEOGRAPHIC DISTRIBUTION The grapefruit is thought to be an accidental hybrid of the orange and pumello. It was first discovered growing in the Caribbean in the 1700s. Seeds were brought to Florida in 1823. By the 1880s, Florida growers were shipping fruit to northern cities.

TREE DESCRIPTION The grapefruit is a medium evergreen. A mature tree may exceed 35 feet in height. The canopy tends to be more spreading and open than the canopies of other varieties of *Citrus*. Occasional thorns are present on smaller branches. The dark green leaves are ovate, measuring between 3 and 5 inches in length. Petioles are prominently winged. The white flowers have 4 petals and measure nearly 2 inches in diameter.

FRUIT CHARACTERISTICS The fruit of the grapefruit, like that of other *Citrus*, is an atypical berry or hesperidium. It is round to faintly pyriform, measuring between 3 1/2 and 6 1/2 inches in diameter. Skin color ranges from pale yellow to yellowish green. Some cultivars develop a pink blush. The pulp is formed into 11 to 14 segments. Depending on the cultivar, the flesh may be seedless or may contain more than a dozen seeds.

CULTIVARS Grapefruit cultivars recommended for dooryard planting in Florida include 'Redblush,' a red-fleshed variety; 'Marsh,' a nearly seedless, yellow-fleshed cultivar; 'Thompson' a nearly seedless, pink-fleshed cultivar; 'Flame' a seedless red-fleshed cultivar; 'Sweetie' a hybrid between the grapefruit and pummelo; 'Rio Red,' a productive red-fleshed variety; 'Star Ruby,' a red-fleshed varity; 'Duncan' a pale yellow-fleshed variety with excellent flavor, but a high seed content; 'Foster Pink,' a pink-fleshed cultivar with a high seed content; and 'Burgundy,' a late-season, nearly seedless cultivar with red flesh.

RELATIVES The *Citrus* genus belongs to the Rutaceae family. Other *Citrus* species discussed within this book include the lime, *Citrus* spp.; the lemon, *Citrus limon;* the mandarin, *Citrus reticulata;*

and the orange, *Citrus sinensis.* The pummelo or shaddock, *Citrus maxima,* is a progenitor of the grapefruit. It produces the largest fruit of any citrus species. The fruit can attain the diameter of a soccer ball. The flavor is thought by many to exceed that of the grapefruit. Unfortunately, the thick rind consumes much of the interior, resulting in a low flesh to weight ratio. The pummelo makes a fine dooryard tree for south Florida. Recommended cultivars include 'Chandler,' 'Kao Pan,' 'Nakhon,' 'Ogami,' 'Reinking,' and 'Tresca.'

CLIMATE The grapefruit is marginally less cold hardy than the orange. It is commonly grown as far north as the Cape Canaveral area on the east coast, and the Tampa Bay region on the west coast. The tree suffers serious damage when the temperature drops to 26° F.

CULTIVATION Once established, the grapefruit has few cultural requirements. Many abandoned trees have lived on for decades and have continued to bear fruit long after forest hammocks have grown up around them. The tree will adapt to a wide range of soil types. Fertilization and irrigation requirements closely correspond with those of the sweet orange. The grapefruit suffers from many of the same pests and ailments that affect other types of *Citrus*. It is especially vulnerable to citrus canker. Other common diseases include algal leaf spot, greasy spot, and root rot. The grapefruit can be propagated with budding and grafting techniques commonly applied to other *Citrus*.

HARVEST AND USE In Florida, fruit usually peak in flavor from Thanksgiving through January. However, the fruit will hold on the tree for several months without any appreciable loss in quality. This presents a great advantage to the homeowner. By May, the fruit begins to lose flavor and the seeds begin to sprout within the flesh. Fruit will store at room temperature for up to a week and will keep in the refrigerator for up to 3 weeks.

Compare the grapefruit (left) with its parent, the pumello (right).

Grumichama

SCIENTIFIC NAME: *Eugenia brasiliensis*
FAMILY: Myrtaceae
OTHER COMMON NAMES: Brazil Cherry, Grumixama

Fruiting Calendar

JAN	FEB	MAR	APR	MAY	JUN	JUL	AUG	SEP	OCT	NOV	DEC

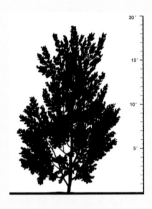

The grumichama is a small, attractive tree that produces a delicious, cherrylike fruit. Indeed, many regard this species as the premier member of the fruit-rich *Eugenia* genus. Although the fruit is small and is somewhat susceptible to attack by the Caribbean fruit fly, the flavor is outstanding. For south Florida residents who yearn for a fruit resembling the northern sweet cherry, the grumichama represents an excellent choice.

Characteristics

Overall Rating	★★★★
Ease of Care	★★★★
Taste/Quality	★★★★
Productivity	★★★★
Landscape Value	★★★★
Wind Tolerance	★★★★
Salt Tolerance	★★
Drought Tolerance	★★
Flood Tolerance	★★★
Cold Tolerance	★★

Known Hazards

None

GEOGRAPHIC DISTRIBUTION The grumichama is native to southern Brazil. It is cultivated to some extent in Australia, Hawaii, Jamaica, Costa Rica, and the Philippines. It has been grown in Florida since at least 1911, but has never achieved the popularity that it deserves.

TREE DESCRIPTION The grumichama is a small evergreen forming a compact tree or multi-trunk shrub of 15 to 25 feet in height. The foliage is glossy, thick, and ornamental. Leaves are opposite, deep green, oval, entire, and leathery, with slightly recurved margins. They measure from 3 to 6 inches in length. Emerging growth is a brilliant, cranberry-red. Leaves remain on the tree for about 2 years before yellowing and dropping.

The grumichama has a moderate rate of growth, gaining about a foot per year. Fruit production begins when the plant reaches 4 or 5 feet in height. The tree makes an outstanding landscape specimen. It can be used as a thick hedge, although no trimming should be undertaken until after the main crop has set. The flowers—small, white and showy—are borne on long, thin pedicils, which emerge from the leaf axils. The flowers have 4 petals, numerous stamens, and measure about 1 inch in diameter.

FRUIT CHARACTERISTICS The fruit is an oblate berry with persistent sepals. It usually measures just under an inch in diameter. At least one selection produces a fruit that exceeds 1 1/2 inches in diameter. Fruit are born singly on new growth. As the fruit matures it turns from bright green to red to purple-black. The skin is glossy and thin. The flesh is firm but juicy and ranges in color from off-white to red. The flavor is sweet, aromatic, and balsam-like. The pulp encloses between 1 and 3 hard, gray-green seeds, grouped toward the center of the fruit.

Myrtaceae family discussed within this book include the blue grape, *Myrciaria vexator;* cattley guava, *Psidium cattleianum;* cherry of the Rio Grande, *Eugenia aggregata;* feijoa, *Feijoa sellowiana;* guava, *Psidium guajava;* and jaboticaba, *Myrciaria* spp.

CLIMATE The grumichama requires a subtropical climate. It has survived temperature drops to 25° F, although the foliage is damaged at about 27° or 28° F. It can grow in protected locations as far north as Cape Canaveral on the east coast and Pinellas County on the west coast.

CULTIVATION The grumichama is a low-maintenance tree. It prospers in the sandy soil found across much of the Florida peninsula, but makes poor growth on oolitic limestone. It prefers a slightly acid growing medium. The tree will endure full sun or partial shade. The grumichama requires regular irrigation when young. Once established, it is moderately drought tolerant. Few pests and diseases affect the grumichama in Florida. Maggots of the Caribbean fruit fly sometimes infest the fruit, and no practical control exists. A pest new to Florida, the lobate lac scale, may attack the foliage. Birds sometimes raid the fruit. The grumichama is usually grown from seed. Grafting is rarely practiced as plants sucker freely from beneath the graft union. Selected cultivars are reproduced through leafy cuttings.

HARVEST AND USE In Florida, the grumichama typically ripens in April and May, although sporadic off-season fruiting sometimes occurs. The main crop ripens over the course of about 2 weeks. To minimize the potential for fruit fly infestation, the fruit should be picked as soon as it reaches full coloration. The fruit is most often eaten out of hand. It also makes excellent jellies and preserves.

The sweet flesh of the grumichama contains one or two large seeds.

CULTIVARS Few cultivars are available in Florida, although some nurseries have selected superior clones. Most nursery stock is grown from seed.

RELATIVES Members of the genus *Eugenia* discussed within this book include the cherry of the Rio Grande, *Eugenia aggregata;* and pitomba, *Eugenia luschnanthiana.* Members of the

Mature trees are capable of producing heavy crops.

Guava

SCIENTIFIC NAME: *Psidium guajava*
FAMILY: Myrtaceae
OTHER COMMON NAME: Guayaba (Spanish)

Fruiting Calendar

JAN	FEB	MAR	APR	MAY	JUN	JUL	AUG	SEP	OCT	NOV	DEC

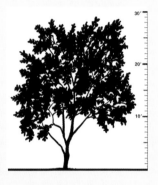

Characteristics

Overall Rating	★★★
Ease of Care	★★★★★
Taste/Quality	★★★★
Productivity	★★★★★
Landscape Value	★★★
Wind Tolerance	★★
Salt Tolerance	★★
Drought Tolerance	★★★★
Flood Tolerance	★★★★
Cold Tolerance	★

The common guava is easy to grow and tolerates adverse conditions. The tree is attractive and the patchy bark lends visual interest. It bears prolific quantities of fruit. The aromatic fruit is well regarded throughout the American tropics and beyond. Unfortunately, the guava is a primary host of the Caribbean fruit fly. In addition, it is classified as an invasive exotic. When making planting decisions, the gardener should weigh these two drawbacks against the guava's many attributes.

Known Hazards

The plant is an invasive exotic.

GEOGRAPHIC DISTRIBUTION The guava originated in the American tropics, probably in Central America. It was widely distributed prior to the arrival of the Europeans.

The guava is an invasive exotic and grows wild in many parts of western Miami-Dade, Broward, and Palm Beach Counties. Naturalized populations are found as far north as Indian River County and Pinellas County. The gardener should carefully consider the ramifications before planting this tree in areas near sensitive native habitats.

TREE DESCRIPTION The guava is a small, ever-green tree, attaining a height of between 20 and 30 feet. The canopy is spreading, somewhat rounded but irregular. The texture of the foliage is coarse. The medium-green leaves are opposite, oblong-elliptic to oval. They measure between 3 and 7 inches in length. Emerging growth may have a pink or reddish tint. The trunk is short and low branch-ing. The reddish-brown bark is smooth and patchy, flaking off to reveal white, green, and tan under lay-ers. The guava has a rapid rate of growth. White flowers, about an inch in diameter, are born prolif-ically on new growth. They have a conspicuous tuft of stamens and 4 or 5 petals.

The flower is typical of that produced by many members
of the Myrtaceae family.

FRUIT CHARACTERISTICS The fruit of the guava is a large berry, usually measuring between 2 and 5 inches in diameter. It may be globose, ovoid, or

Fruit fly damage to the interior of the fruit is common.

pyriform. The sepals remain attached to the apex. The light-green skin is smooth, often slightly undu-lating or bumpy. The flesh is thick and somewhat granular. It is crisp at first, but softens as the fruit ripens. In taste, it is subacid, juicy, and aromatic. Pulp toward the center of the fruit contains many small seeds.

CULTIVARS Because the fruit must be bagged to protect it from the Caribbean fruit fly, preference should be given to large-fruited cultivars. Recommended cultivars include 'Thai Giant,' 'White Indian,' 'Hong Kong Pink,' 'White Indonesian,' 'Lucknow 49,' 'Red Indian,' 'Detwiler,' 'Ruby X Supreme,' 'Peruvian White,' and 'Supreme.'

RELATIVES This book profiles several members of the large Myrtaceae family, including the blue grape, *Myrciaria vexator;* cattley guava, *Psidium cattleianum;* cherry of the Rio Grande, *Eugenia aggregata;* feijoa, *Feijoa sellowiana;* grumichama, *Eugenia braziliensis;* jaboticaba, *Myrciaria* spp.; pitomba, *Eugenia luschnanthiana;* and stoppers, *Eugenia* spp. Other members of the *Psidium* genus are described in the section pertaining to relatives of the cattley guava.

CLIMATE The guava can only survive a few degrees of frost and may be greatly damaged at 29° F. However, even when frozen to the ground it will usually resprout from the roots.

CULTIVATION The guava is a low-maintenance tree. It will grow in a wide range of soils, including those with a pH ranging from 4.5 to 8.5. It can withstand considerable drought. It can also tolerate short periods of flooding and will grow on soils too moist for most fruit trees.

The guava is highly susceptible to attack by the Caribbean fruit fly. Only two effective controls exist. The first is to bag individual fruit. The second is to force flowering and fruiting during the winter when fruit fly populations are low. If the tree is pruned back in October, it will bear fruit in January and February. Superior clones can be propagated through air layering and by rooting cuttings under intermittent mist. Grafting is rarely practiced, as grafted plants will produce shoots beneath the graft union.

HARVEST AND USE In Florida, production peaks in the late summer and fall. The tree is capable of prodigious yields. The fruit should be picked as soon as it begins to change color. It deteriorates quickly if stored at room temperature. The fruit is superb when eaten fresh. Seeds are generally con-sumed with the flesh. However, seeds are strained when the guava is used as an ingredient in sauces, pastes, chutneys, puddings, pies, milkshakes, and ice cream.

Imbe

SCIENTIFIC NAME: *Garcinia livingstonei*
FAMILY: Clusiaceae

Fruiting Calendar

JAN	FEB	MAR	APR	MAY	JUN	JUL	AUG	SEP	OCT	NOV	DEC

Characteristics

Overall Rating	★★
Ease of Care	★★★★
Taste/Quality	★★
Productivity	★★★
Landscape Value	★★★★
Wind Tolerance	★★★
Salt Tolerance	★★
Drought Tolerance	★★★★
Flood Tolerance	★★★
Cold Tolerance	★

The imbe is a species of minor import that is occasionally grown as a curiosity in south Florida. It is a relative of the famous mangosteen, which bears a fruit that is considered among the best in the world. The fruit produced by the imbe is not among the world's best, but is pleasant and is readily enjoyed by most. The small tree is worth growing both for its foliage and for its fruit in warmer areas of the state.

Known Hazards

None

GEOGRAPHIC DISTRIBUTION The imbe is native to tropical East Africa. It is rarely planted in Florida outside of tropical fruit tree collections and arboretums.

TREE DESCRIPTION The imbe is an evergreen shrub or small tree, attaining a height of 10 to 20 feet in Florida. The stiff, leathery leaves measure 4 to 6 inches in length. Veins are much lighter than the surrounding deep-green blade, and this contrast makes the foliage ornamental. Branches grow at a 90-degree angle from the trunk. The tree is slow growing, usually requiring 5 years or more to come into production. The imbe is dioecious. A male and female tree are required to ensure adequate fruit set. Flowering takes place in late April and May.

FRUIT CHARACTERISTICS The fruit are oval to globose, measuring from 1 1/2 to 2 inches in length. The skin is orange to reddish-orange in color. Beneath the skin is a thin layer of orange, juicy pulp. In flavor, it is sweet and tangy, somewhat resembling an apricot. Embedded within the pulp are 1 or 2 large seeds.

The fruit, like the foliage, is decorative. While the flavor is good, the imbe does not rank among the elite of the *Garcinia* genus.

CLIMATE The imbe is not purely tropical—unlike its glamorous cousin, the mangosteen. It suffers leaf damage at 30° F, but will withstand temperature drops to about 26° F. It is probably not worth attempting north of Ft. Pierce, on the east coast, or Sarasota, on the west coast.

RELATIVES The only other member of the family Clusiaceae profiled within these pages is the cherapu or button mangosteen, *Garcinia prainiana.* Several *Garcinia* species from Asia are described within the subsection pertaining to relatives of cher-

apu. Many Clusiaceae species from the American tropics were formerly classified within the genus *Reedia,* but are now grouped within the genus *Garcinia.* Among these are such fruiting species as the bakupari, *Garcinia brasiliensis,* of southern Brazil and Paraguay; the charchuela, *Garcinia macrophylla,* of Brazil; the Cuban mangosteen, *Garcinia aristata,* of the Caribbean; the madrono, *Garcinia madruno,* of Central America; and the mameyito, *Garcinia edulis,* of Central America. The distantly related mamey apple, *Mammea americana,* is native to the Caribbean and northern South America. The tree is occasionally planted in south Florida, where it is grown primarily for its showy flowers and magnolialike foliage. The fruit is of fair but not outstanding quality, tasting something like a washed-out mango.

The madrono is a tropical American cousin of the imbe.

CULTIVATION The imbe is a hardy tree and requires only moderate care. It prefers lightly acidic soil. Once established, the imbe is drought resistant, although fruit production is enhanced if irrigation is provided during spring dry spells. The tree is somewhat salt tolerant. The imbe prefers full sun but will tolerate some shade. Although the imbe has few serious pests and diseases, the Caribbean fruit fly occasionally attacks the fruit. The imbe is generally grown from seed.

HARVEST AND USE The fruit matures in June. It is picked when fully colored. Storage characteristics are poor, as the fruit has thin skin and is easily bruised. It is most often eaten out of hand, but can also be used to make jelly or as a flavor for smoothies, milkshakes, and ice cream.

Jaboticaba

SCIENTIFIC NAME: *Myrciaria* spp.
FAMILY: Myrtaceae

Fruiting Calendar

JAN	FEB	MAR	APR	MAY	JUN	JUL	AUG	SEP	OCT	NOV	DEC

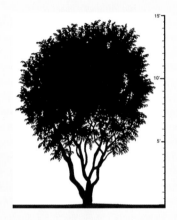

Characteristics

Overall Rating	★★★★★
Ease of Care	★★★
Taste/Quality	★★★★★
Productivity	★★★★★
Landscape Value	★★★★★
Wind Tolerance	★★★
Salt Tolerance	★
Drought Tolerance	★
Flood Tolerance	★★★★
Cold Tolerance	★★★

The jaboticaba ranks as one of the most desirable species for dooryard planting, surpassing citrus in many respects. It is a sensational landscape specimen, with peeling bark and a dense crown of lacy foliage. The fruit deserves equal praise. Aficionados consider the flavor to be the finest of any berry-type fruit. The tree bears profusely and repeatedly. The jaboticaba is highly recommended for planting in south Florida and protected areas of central Florida.

Known Hazards

Fruit skins contain tannins and should not be consumed in quantity

GEOGRAPHIC DISTRIBUTION The jaboticaba is native to the coastal forests and hilly regions of southern and central Brazil. It is also present in adjacent areas of Paraguay, Uruguay, and northern Argentina. It has not been distributed as widely as many subtropical species of lesser merit. The jaboticaba has been grown in Florida since at least 1908. Despite its tremendous promise as a dooryard crop, no serious attempt has been made at improvement. With the spread of citrus canker and the arrival of serious insect pests that do not affect this tree, the future of the jaboticaba is bright.

TREE DESCRIPTION The jaboticaba is a small, evergreen tree with a rounded, symmetrical crown of fine texture. In its native habitat, the tree reaches a height and spread of nearly 40 feet. In Florida, mature specimens average 15 to 20 feet in height. Leaves are lanceolate, opposite, measuring between 1 and 4 inches in length. The tips of emerging foliage often have an attractive pink or reddish tint.

Emerging growth of the jaboticaba.

As the tree matures, the trunk and major branches take on a gnarled, ancient appearance. The jaboticaba ordinarily grows from a single trunk. However, it is not unusual for specimens to have multiple trunks, which may fuse together. The tree branches close to the ground. The branches are profuse, angling upward and outward to form a bushy canopy. The bark is smooth and flakes off in small plates to form multicolored patches. The jaboticaba is shallow rooted and feeder roots form a dense mat of fine fibers.

The jaboticaba makes an ornamental landscape specimen and lends itself to many uses. It can be planted singly. It can also be established as a high hedge, with 6 feet separating the tree centers. When the jaboticaba is planted as a hedge, trimming the foliage does not significantly reduce production, since the fruit is borne on older, interior wood.

Even with faithful care, the jaboticaba is a tree of slow growth. A seedling will usually require 8 or more years before coming into production. A Brazilian proverb focuses on the jaboticaba's slowness to bear:

An elderly man was tending to a small plant with great care and affection. A younger man approached and struck up a conversation.
Isn't that a jaboticaba?
Yes.
How long will it take to bear fruit?
About ten years.
And you expect to be around for the harvest?
No, I am near the end of my life.
Then what advantage can you hope to gain from all your work?
None, except to know that no one would eat jaboticaba if everyone thought as you.

Once fruit production begins it increases each year until maturity, which occurs at about 25 years. The tree may live for more than 150 years. Fruiting is presaged by the appearance of hundreds of tiny polyps—about the size of a pinhead—on the bark of the tree. These swell into small green buds. The buds soon burst forth into clusters of spectacular tufted white flowers. From a distance, the flowers resemble snow gathered on the trunk and branches. Rain and overhead irrigation are harmful to fruit set during the first 3 days after bloom. Two days after they appear, the flowers begin to desiccate. From this point forward, fruit quality depends on the tree receiving a steady supply of water.

A jaboticaba in flower is a sight to behold.

FRUIT CHARACTERISTICS The jaboticaba is cauli-florous. Flowers and fruit are borne on short stalks emerging directly from the trunk and main branches. The tiny green fruit form among the dried flowers and swell rapidly. Between 20 and 30 days after flowering, the fruit reaches full size. It then undergoes a swift color change, with the skin streaking bronze and crimson, turning purple, then black.

The fruit is a globose berry. It may be as small as a pea or larger than a ping-pong ball. It superficially resembles the fruit of a slip-skin grape, with tough skin and whitish, translucent flesh. The pulp contains between 1 and 4 seeds, which vary widely in shape, size and color depending on the species. In terms of eating quality, the jaboticaba is positioned among the elite of subtropical fruit. The flavor is superb. The pulp is sweet, aromatic, and rich, with hints of balsam and cedar. The texture is softer and more gelatinous than that of a grape.

The fruit can become quite large. Whatever the size, a ripe jaboticaba is always an exceptional treat.

SPECIES AND STRAINS The taxonomy of the jaboticaba has long been a source of confusion. The common name is presently applied to about 10 species composed of more than 25 recognizable strains. Brazil's Federal University at Viçosa maintains an extensive germplasm collection. While some overlap is probable, species identified in the literature include *Myrciaria aurean, Myrciaria baporite, Myrciaria cauliflora, Myrciaria jaboticaba, Myrciaria coronata, Myrciaria grandiflora, Myrciaria oblongata, Myrciari phitrantha, Myrciaria spiritosantensis, Myrciaria tenella,* and *Myrciaria trunciflora.* Many strains have been described, including 'Branca,' the rare and well-regarded white jaboticaba; 'Coroa,' a small-fruited type; 'Coronado,' a jaboticaba of excellent flavor and eating quality, 'De cabinho' which bears small, pink fruit on elongated stalks, 'De cipo,' a rare variety that produces fruit toward the branch tips; 'Jabotica-tuba,' a variety that produces very large

fruit of very good flavor; 'Murta,' a variety that bears small to medium fruit of good quality; 'Olho do Boi,' which bears enormous fruit of indifferent quality; 'Paulista,' which produces large, tough-skinned fruit of very good quality; 'Ponhema' which produces a leathery skinned fruit with a pointed apex of fair quality; 'Rajada,' which produces a bronze-skinned fruit of very good quality; and 'Rujada,' which bears variegated fruit. 'Sabará' is the most famous variety. It was named after a historical mining city, known today for its annual jaboticaba festival. The fruit are medium-size with dark purple to black coloration. The pulp is sweet and aromatic.

Several strains are available in Florida. Most nurseries carry a small-leaved variety that has been grown in Florida for at least 75 years. The tree typically reaches no more than 15 feet in height. The shiny, black fruit is of excellent quality and is borne in great profusion. A mature tree may bear up to 6 crops annually. A large-leafed variety that bears a large, thin-skinned fruit of excellent quality is occasionally available. It produces only one crop annually. Another large-leafed strain, with a stout and sparse habit of growth, produces a thick-skinned fruit of fair quality. A rare 'weeping' strain produces a small fruit of good quality, but has extremely high water needs.

The small-leaf jaboticaba commonly sold by Florida nurseries.

A large-leaf strain of jaboticaba that is occasionally available in Florida.

RELATIVES Other Myrtaceae species discussed within this book include the blue grape, *Myrciaria vexator;* cattley guava, *Psidium cattleianum;* cherry of the Rio Grande, *Eugenia aggregata;* feijoa, *Feijoa sellowiana;* grumichama, *Eugenia braziliensis;* guava, *Psidium guajava;* pitomba, *Eugenia luschnanthiana;* and stoppers, *Eugenia* spp.

The genus *Myrciaria* is composed of about 70 species from tropical America and the Caribbean. The yellow jaboticaba, *Myrciaria glomerata,* has large, glossy green leaves and rough bark. Like the jaboticaba, it is cauliflorous, although it also bears fruit on newer growth. The trunk is gray and lightly furrowed. The yellow fruit is tasty but the rind is thick and the flesh scant. The camu camu, *Myrciaria dubia,* is native to the Amazon basin. It bears small, dark red fruit with shiny skin. The pulp is juicy and acidic and contains high levels of Vitamin C. The rumberry, *Myrciaria floribunda* is a minor but well-regarded fruiting species, native to the West Indies and northern South America. The tree is compact, handsome, drought tolerant, hurricane resistant, and well behaved. The fruit, which forms on new growth, is diminutive but aromatic and delicious. It is the prime ingredient in guavaberry liqueur, a traditional Christmas drink in the Virgin Islands. This species is worth planting on a wider basis in south Florida.

CULTIVATION The jaboticaba will prosper without fertilization or other care; however, it must receive an adequate supply of water. If the top 4 inches of soil are permitted to dry out, leaves begin to wilt and the tree may suffer irreversible injury. The jaboticaba thrives if it is watered 2, 3, or even 4 times a week. It survives occasional river flooding in its native habitat. While the foliage of the jaboticaba is wind resistant, the tree will blow over when winds exceed 100 miles per hour. It can usually be righted without any lasting damage.

The jaboticaba is sensitive to over-fertilization and prefers slow-release formulas. Some growers apply frequent, light applications of high-quality palm fertilizer to the adult tree. The tree should be pruned to facilitate harvest. Unless the lower canopy is thinned, it becomes difficult to pick fruit from interior branches. In Brazil horizontal branches lower than 3 feet are eliminated and the tree is pruned to form a vase shape.

PESTS AND DISEASES Few pests or diseases affect the jaboticaba in Florida. Rust occasionally attacks the fruit and flowers. Raccoons, opossums, and squirrels sometimes develop an affinity for the fruit. Fortunately, the fruit's tough skin prevents infestation by the Caribbean fruit fly. In Brazil, birds are by far the most vexatious pests. Reportedly, they become so intent on the feast that they remain completely silent: the only sound emitted is a gentle rustling caused by the fall of hundreds of empty rinds. In Florida, grackles occasionally steal the fruit; however, Florida birds apparently lack the degree of rapaciousness found in their cousins to the south.

PROPAGATION The jaboticaba is usually grown from seed. Seeds germinate in about 35 or 40 days, and send up rust-colored filaments, which spit at the tip, forming tiny leaves. Some success has been obtained reproducing jaboticaba through leafy cuttings. Air layering is sometimes successful, although roots are slow to form.

HARVEST AND USE In Florida, the main crop usually ripens in March and April and may be stimulated by heavy rain. Many trees fruit sporadically throughout the year. A mature tree may produce more than 100 pounds of fruit over the course of a season. The fruit should be harvested within a few days after it reaches full coloration. Once the fruit has been picked, deterioration accelerates. At room temperature the fruit spoils within 2 days.

When eating the fruit directly from the tree, the best technique is to nip the skin with the teeth and, using the thumb and index finger, squeeze the pulp into the mouth. The empty skin is then discarded. The seeds may be separated in the mouth, although some simply swallow the pulp, seeds and all. The jaboticaba makes a fine jelly and has also been used in the production of wines and brandies.

The jaboticaba is capable of prodigious yields.

Jackfruit

SCIENTIFIC NAME: *Artocarpus heterophyllus*
FAMILY: Moraceae
OTHER COMMON NAME: Jaca (Spanish)

Fruiting Calendar

JAN	FEB	MAR	APR	MAY	JUN	JUL	AUG	SEP	OCT	NOV	DEC

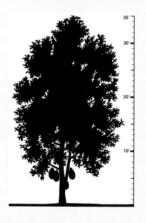

Characteristics

Overall Rating	★★★★★
Ease of Care	★★★★
Taste/Quality	★★★★
Productivity	★★★
Landscape Value	★★★★★
Wind Tolerance	★★★
Salt Tolerance	★★★
Drought Tolerance	★★★
Flood Tolerance	★★★
Cold Tolerance	★

The jackfruit is the world's largest tree-borne fruit. The fruit can take on the proportions of a large watermelon and can weigh more than 80 pounds. Oblong, spiky, oddly reticulated, the fruit dangle from the trunk and major branches on stout cords. A fruiting tree presents a sensational visual impact. The fruit possesses a fine tropical flavor. The tree is stately and handsome and serves as a beautiful shade tree when it reaches maturity. It deserves to be planted on a broad scale and its popularity as a dooryard tree in Florida is soaring.

Known Hazards
Raw seeds are indigestible and contain a trypsin inhibitor, which is dispelled through cooking.

GEOGRAPHIC DISTRIBUTION The jackfruit is native to southern India. Its origin has generally been traced to the rain forests of the Western Ghats, south of Bombay. The jackfruit is an important crop throughout Southeast Asia. The species has been grown in Florida since at least 1886, but languished in obscurity for almost a century. The introduction of superior cultivars has spurred recent interest.

TREE DESCRIPTION The jackfruit is a medium to large evergreen tree. In Florida, it rarely exceeds 40 feet. The tree tends to be upright when young, but spreading with age. Leaves are oval or elliptic in form, 5 to 8 inches in length. They are dark green above, light green below, with a glossy sheen on the upper surface. When cut or otherwise damaged, all parts of the tree exude white latex. The jackfruit is an attractive landscape specimen and has a fast rate of growth.

The female (left) and male (right) flower heads appear quite different when placed side by side. Once the male head has dispensed its pollen it rapidly decays and drops from the tree.

Few trees exceed the jackfruit as a landscape specimen. This five-year-old 'NS1' seedling produces exceptionally heavy crops.

The tree is monoecious; it produces both male and female flower heads. Male flowers are born on thin stalks, which usually emerge from new growth. Once they have produced pollen, the male flowers wither and are consumed by mold. Female flowers are born on thicker shoots, which usually emerge from older wood. A caplike receptacle supports the base of each female flower. The jackfruit can be hand pollinated by lightly rubbing the surface of a female fruit with a pollen-laden male flower.

FRUIT CHARACTERISTICS The compound fruit can measure more than 2 feet in length. Typical specimens weight from 10 to 50 pounds. The exterior is covered with several thousand, closely packed pyramidal spines or studs. The rind is leathery and tough. The ripe fruit emits a strong, sweet odor, objectionable to some, relished by others.

Within the rind lie the perianths. These fleshy bulbs—the edible portion of the fruit—each surround a single seed. The perianths are embedded within a mass of tough, linguini-like fibers known as the "rag." Flesh color ranges from light yellow to orangy-red. A pithy core runs through the center of the fruit. The flavor of the ripe fruit resembles a banana crossed with a mango and pineapple. It is sweet, aromatic, and delicious. In some varieties the flesh is crisp and crunchy; in others it is soft and fibrous.

The seeds are contained in translucent, inedible sacks in a pocket at the center of each perianth. The seeds measure about an inch in diameter. They are tan, glossy, and enclose a creamy white interior. An individual fruit may contain between 35 and 350 seeds.

CULTIVARS Desirable characteristics in a jackfruit cultivar include high recoverable flesh ratio, low seed count, small- to medium-size fruit, low latex, heavy production, minimal rag, crisp texture, and outstanding flavor. Several cultivars recommended for Florida are described below.

'Borneo Red'–Chris Rollins, curator of Miami-Dade County's Fruit and Spice Park, introduced this cultivar to Florida. Fruit quality is outstanding. The flavor is rich and sweet, with overtones of melon. The flesh is crisp and firm, orangy-pink in color. Fruit weigh an average of about 20 pounds.

'Mai I' - This Vietnamese cultivar is a top choice for dooryard planting. Richard Wilson, proprietor of Excalibur nursery, brought this variety to Florida. The tree is upright and moderately vigorous. Production is moderate. Fruit quality is outstanding, with low latex, little rag, low seed count, thick-fleshed bulbs, and a high flesh recovery ratio. Fruit average about 15 pounds in weight.

'NS1' - This was one of the first superior jackfruit clones brought to Florida, and is still among the best. Rare-fruit pioneer William Whitman introduced this cultivar in 1986. NS stands for the Negeri Sembilan province of Malasia, where the tree is thought to have originated. The tree is vigorous and productive. The fruit, which weigh about 10 pounds, have a low seed count and thick, orange-fleshed bulbs.

'J-31' - This Malaysian cultivar produces fruit of superior quality. Production is moderate. The flesh is crunchy and rich. The fruit boasts a high flesh recovery ratio. Fruit weigh about 25 pounds on average.

'Dang Rasimi' - This productive Thai cultivar is a good choice for the home garden. While the seed count is high, the flesh is crisp and the quality is outstanding. Fruit average about 17 pounds.

'Cheena' - This Australian cultivar is actually a cross between the jackfruit and its close relative, the champedak. Cheena has a low seed count and a high flesh recovery ratio. The internal arrangement of the fruit makes for easy preparation. Although the fruit has soft flesh, the flavor is rich and sweet. The average fruit weighs about 5 pounds.

Other varieties that are worth planting include 'Black Gold,' a productive Australian cultivar with soft, deep orange flesh; 'Cochin,' an Australian cultivar that produces small, round fruit of exceptional quality; 'Golden Nugget,' an Australian cultivar with a low seed count and high flesh recovery ratio; 'Golden Pillow,' a Thai cultivar with a low seed count and thick, mild flesh; 'Honey Gold,' an Australian cultivar that produces small, flavorful fruit; and 'J-30,' a productive Malaysian cultivar with outstanding flesh quality.

RELATIVES The jackfruit is a member of the Moraceae family, and is distantly affiliated with the fig, *Ficus carica;* mulberry, *Morus* spp.; and che, *Cudrania tricuspidata,* reviewed within separate sections of this book.

The genus *Artocarpus* includes approximately 50 species native to the Indian subcontinent, Southeast Asia, and various islands of the Pacific.

The breadfruit, *Artocarpus altilis,* originally from the South Pacific, is a staple crop in many tropical regions and will grow in the Florida Keys. The starchy fruit resembles a small, round, smooth-skinned jackfruit. The kwai muk, *Artocarpus hypargyraeus,* is native to China. It is occasionally planted in Florida as an ornamental. The fruit is about 2 inches wide and contains edible reddish pulp. The tree is harmed at 28° F and killed at 25° F. The lakoocha or monkey jack, *Artocarpus lakoocha,* is native to the sub-Himalayan regions of India. The fruit is dull yellow with sometimes with a pink tinge. The sweet-sour pulp may be eaten fresh or made into chutney.

Several notable *Artocarpus* species require a more tropical climate than that found in peninsular Florida. The champedak, *Artocarpus integer,* native to Malaysia, resembles the jackfruit but is smaller and somewhat elongated. It is adversely affected by winter cold in Florida. Specimens that have survived several winters have failed to fruit. Other ultra-tropical species include the marang, *Artocarpus odoratissimus,* of Borneo, which produces a 6-inch fruit with white, juicy flesh and outstanding flavor; the pedalai, *Artocarpus sarawakensis,* of northeastern Borneo, which produces a 6-inch, orange-skinned fruit equal in flavor to that of the marang; the pingan, *Artocarpus sericicarpus,* of Borneo, which produces a fruit similar to that of the pedalai; and the entawak, *Artocarpus ansiophyllus,* of Borneo and Sumatra, which bears a 4-inch, brownish-yellow fruit with orange-red flesh.

CLIMATE The jackfruit requires a tropical or near-tropical climate. The tree will suffer leaf burn and browning as a result of a light freeze. A well-established jackfruit can survive a temperature drop to about 28° F, although it will suffer defoliation. Lower temperatures cause limb damage. The tree is resilient and a mature specimen may spring back from temperatures as low as 26° F. Young trees are especially susceptible to cold injury.

CULTIVATION The jackfruit is a low-maintenance tree. It prefers deep, nutrient-rich soil, but makes acceptable growth on sandy soil and on the oolitic limestone soils of Miami-Dade County. The jackfruit has moderate salt tolerance. In Thailand it is planted as a barrier behind coconut palms to protect orchards of other fruit trees from saltwater intrusion. The jackfruit prefers full sun, but will grow and fruit in partial shade.

Proper drainage is critical, as saturated soil can lead to root rot. The young tree should be watered twice a week until it is established. Irrigation should be discontinued over the late fall and early winter, even though the tree's growth is more or less continuous. A winter rest period seems to stimulate fruiting and vigorous spring growth. The jackfruit should receive frequent, light applications of a balanced fertilizer.

Pruning is required to control tree size. Terminal buds should be frequently headed to encourage additional branching. The goal is to form a short, thickly branched, heavily foliated tree that can be kept at a height of between 10 and 15 feet. The tree should not be allowed to fruit until it has reached at least 8 feet in height.

The jackfruit sometimes fruits near the base of the trunk.

PESTS AND DISEASES Very few pests or diseases bother the jackfruit in Florida.

PROPAGATION Jackfruit grown from seed usually produce acceptable fruit. The first generation of seedlings appears to retain about 90 percent of the characteristics of the parent. Seedlings tend to be more vigorous than their grafted counterparts. Seeds quickly lose viability once removed from the fruit. They germinate in 2 to 3 weeks. Considerable advances have been made in grafting jackfruit. The favored technique is a modified veneer graft. Cambium-to-cambium contact must be obtained along a surface exceeding 3 1/2 inches in length. The grafted tree must be kept in a humid environ-

ment until the graft takes. The tree should be kept under shade cloth until it is about 3 feet tall. Jackfruit can also be cloned through air layering and leafy cuttings, although these methods are rarely employed in Florida.

HARVEST AND USE Fruit maturity can be adduced from several factors. In some cultivars the spines flatten and spread. The rind may undergo a subtle color change, from dark green to olive. In all cultivars, the ripe fruit make a dull, hollow sound when tapped with the open palm. As ripening progresses, the fruit begins to exude a sweet odor.

Cleaning a large jackfruit can be a time-consuming task. Some cultivars leak copious amounts of latex when cut. Hands and knives must be coated with vegetable oil so that the latex does not congeal on fingers and blades. In Malaysia, where the jackfruit is popular, there is a time-worn saying: One who eats the jackfruit will be touched by the sap (Siapa makan nangka, dia kena getah). In other words, people must take responsiblity for their own conduct.

The fruit is first cut longitudinally. The central core is removed and discarded. The rind is then turned inside out, so that the bulbs separate from the rag and from each other. Individual bulbs are separated from the rind. The seed and the translucent membrane surrounding the seed must be removed from each bulb. The bulbs can then be eaten, stored, dried, or processed. Cleaned jackfruit will store for 2 weeks in the refrigerator if kept in a sealed container.

The fruit can be eaten fresh or used in milkshakes. The bulbs can be battered and fried or stuffed with various ingredients. The immature fruit is a popular vegetable in some Asian countries. Once the clear seed coat has been removed, the seeds can be boiled or roasted and salted.

The interior arrangement of the jackfruit, as it appears after removal of the pithy core.

This jackfruit, on display at a Singapore market, measured 4 feet long and weighed just under 80 pounds (36 kilograms).

Jamaica Cherry

SCIENTIFIC NAME: *Muntingia calabura*
FAMILY: Elaeocarpaceae
OTHER COMMON NAME: Strawberry Tree, Capulina
(Spanish)

Fruiting Calendar

JAN	FEB	MAR	APR	MAY	JUN	JUL	AUG	SEP	OCT	NOV	DEC

Characteristics

Overall Rating	★★
Ease of Care	★★★★
Taste/Quality	★★★
Productivity	★★★
Landscape Value	★★
Wind Tolerance	★★
Salt Tolerance	★
Drought Tolerance	★★★
Flood Tolerance	★★★
Cold Tolerance	★★

The Jamaica cherry bears abundant quantities of a sweet, cherrylike fruit. It has the advantage of cropping throughout much of the year. The tree is somewhat weedy and soft wooded but is reasonably attractive. It is fast growing and serves as an effective screen. The fruit, while tasty, is almost too small to be taken seriously and is somewhat susceptible to infestation by the Caribbean fruit fly.

Known Hazards

Falling fruit and bird droppings containing fruit remnants can stain walks and patios.

GEOGRAPHIC DISTRIBUTION The Jamaica cherry is native to southern Mexico, Central America, and portions of the Caribbean. Although the fruit has little or no commercial value, the tree has been widely distributed throughout tropical regions of the world. The Jamaica cherry is not related to the true cherry, nor is it closely related to the myriad of subtropical trees commonly referred to as cherries. The Jamaica cherry has escaped cultivation in south Florida and is considered an invasive exotic in various countries.

TREE DESCRIPTION The Jamaica cherry is a small to medium evergreen tree. It typically has a single trunk and is somewhat slender in form with a pyramidal or irregular canopy. It may attain a height of about 30 feet. The branches extend horizontally starting close to the ground, sometimes drooping toward their tips. Leaves are alternate, lanceolate, toothed, and measure from 3 to 5 inches in length. They are soft textured, dark green above, lighter below. The tree is exceptionally fast growing. It may attain a height of 12 feet or more and begin to fruit within 2 years. The wood is soft and the tree has a tendency to break up in high winds. From a landscape perspective, the tree is somewhat coarse textured and is not particularly attractive except when in bloom. Its fast growth makes it valuable as a screen. The bisexual flowers are small but attractive, with white, rarely pink, petals and yellow stamens. They are borne in the leaf axils either singly or in small groups. Although numerous, the flowers are short lived, dropping their petals within hours after they form.

FRUIT DESCRIPTION The fruit is a globose berry, typically measuring about 1/2 inch in diameter. At first, the skin is a bright green. However, it turns light red upon ripening. It is smooth and thin. The flesh is light brown and juicy. The flavor is sweet, musky, figlike, and pleasant. The seeds are yellowish-gray and minute and do not interfere with consumption.

CULTIVARS No known cultivars of Jamaican cherry exist. The tree is uniformly reproduced from seed.

CLIMATE The Jamaica cherry requires near tropical conditions and will not tolerate serious frost. It may succumb to temperatures of 27° or 28° F. The tree has been killed by frost at Ft. Myers and Stuart. However, when not killed outright, it will generally recover quickly.

CULTIVATION The Jamaica cherry will thrive without care. It has no special soil requirements and, indeed, grows well on infertile soil, alkaline soil, and oolitic limestone. It is not salt tolerant and prefers full sun. The only pest of note is the Caribbean fruit fly. An individual tree may be completely overrun or, for unknown reasons, may be free or nearly free of infestation. There is little or no advantage in vegetative propagation. To obtain seeds, the fruit can be lightly mashed and mixed with water, and the seeds allowed to settle. Once separated from the pulp, the seeds are dried and planted at very shallow depth.

HARVEST AND USE The Jamaica cherry is usually eaten out of hand. In Florida, the fruit ripens throughout much of the year, beginning in the spring and ending in the late fall or early winter. The fruit can be picked by hand or can be shaken from branches onto sheets or tarpaulins.

The flower is small and the fruit is tiny, but the flavor is sweet and delicious.

Jujube

SCIENTIFIC NAME: *Ziziphus* spp.
FAMILY: Rhamnaceae
OTHER COMMON NAME: Azufaifo Chino (Spanish)

Fruiting Calendar

JAN	FEB	MAR	APR	MAY	JUN	JUL	AUG	SEP	OCT	NOV	DEC

Characteristics

Overall Rating	★★★
Ease of Care	★★★★
Taste/Quality	★★★
Productivity	★★★
Landscape Value	★★★
Wind Tolerance	★★★
Salt Tolerance	★★★
Drought Tolerance	★★★★
Flood Tolerance	★★★
Cold Tolerance	★★★★

The jujube is a rugged tree. It is drought tolerant, has few pests, and grows well on poor soil. The Chinese jujube can withstand cold winter temperatures and is suitable for growth in north Florida. The semi-tropical Indian jujube is well adapted to south Florida. The datelike fruit borne by both species is delicious. The jujube is a worthwhile addition to the home garden because, even when neglected, it produces heavy crops of good quality fruit.

Known Hazards

Most varieties have sharp spines.

GEOGRAPHIC DISTRIBUTION The Indian jujube, *Ziziphus mauritiana,* is native to southern Asia. It has been distributed to many tropical and subtropical regions and is used as a living fence in some areas. The Chinese jujube, *Ziziphus jujube,* is indigenous to China, where it has been cultivated for at least 4,000 years.

TREE DESCRIPTION The jujube is a small to medium deciduous tree. It may reach a height of 40 feet, but is usually smaller. The leaves are alternate, simple, oblong-elliptic. They measure from 1 1/2 to 3 inches in length. The tree normally has a single trunk, but may form thorny suckers around its base. Branches may have paired spines in the leaf axils. Tiny yellow flowers form in small groups on inch-long inflorescences. Most cultivars are self-fruitful. However, many trees will produce heavier crops with cross-pollination.

FRUIT CHARACTERISTICS The fruit is a dense, fleshy drupe, measuring from 1 to 3 inches in length. The shape is variable. The skin is smooth, thin, and edible. In the Chinese jujube the skin turns from green to mottled red and green to reddish-brown as the fruit matures. In the Indian jujube it changes from green to yellow or orange. The cream-colored flesh is mild, sweet, and somewhat applelike in flavor and texture. The flesh encloses a knobby stone consisting of 2 seeds. The fruit of the Chinese jujube is considered superior to that of the Indian jujube.

CULTIVARS About 500 cultivars have been selected. Cultivars recommended for Florida include 'Li,' an early-season Chinese jujube that bears large round fruit of high quality; 'Sherwood,' a late season Chinese jujube selected in Louisiana; 'Lang,' an early to mid-season Chinese jujube; 'Tigertooth,' a Chinese jujube selected in southeast Alabama that bears datelike fruit of very good quality; 'Silverhill,' a late-season, nearly thornless Chinese jujube suitable for planting in north Florida; and 'Kong Thai,' an Indian jujube selected from Thai seedlings, that is said to fruit well in south Florida.

RELATIVES The Rhamnaceae or Buckthorn family consists of about 55 genera and 900 species. The *Ziziphus* genus consists of about 100 species, several of which produce edible fruit. The Christ-thorn, *Ziziphus spina-christa,* native to semi-arid regions of Africa and the Middle East, bears a commercially viable fruit. The scrub jujube, *Ziziphus celata,* native to central Florida, is protected under the Endangered Species Act. This thorny shrub produces a round, 3/4-inch fruit of uncertain edibility. The Darling plum, *Reynosia septentrionalis,* dis-

cussed within a separate section of this book, is a native tree distantly related to the jujube.

CLIMATE The Chinese jujube, when dormant, can tolerate temperatures below 0° F. The tree requires some winter chill to set fruit, but appears to produce adequate quantities of fruit in south central Florida. The Indian jujube is relegated to the southern half of the peninsula and may be damaged by exposure to a temperature of 26° F.

CULTIVATION The jujube requires little maintenance. It prefers full sun. It will tolerate a wide variety of soils, including alkaline soils. It will also tolerate moderate salinity. The jujube is highly resistant to drought. The tree produces adequate crops without fertilization. Although the tree does not require pruning, any thorny suckers sprouting around the base of the trunk should be removed. No serious pests or diseases bother the jujube in Florida. The tree is sometimes propagated by whip grafting onto root sprouts. Some success has also been obtained at rooting cuttings.

HARVEST AND USE The crop ripens intermittently from July through October. Under favorable circumstances, a tree may produce crops exceeding 100 pounds. Change in color is the primary factor for determining fruit maturity. Some prefer the crisp tartness of the fruit at an intermediate stage. Others prefer the fruit after it has begun to wrinkle and dry. The fruit can be stored for about a week at room temperature. Dried fruit will keep for several months.

An immature fruit dangles from one of the toughest and most versatile trees profiled within these pages.

Kei Apple

SCIENTIFIC NAME: *Dovyalis caffra*
FAMILY: Flacourtiaceae

Fruiting Calendar

JAN	FEB	MAR	APR	MAY	JUN	JUL	AUG	SEP	OCT	NOV	DEC

Characteristics

Overall Rating	★★
Ease of Care	★★★★★
Taste/Quality	★★
Productivity	★★★
Landscape Value	★★★★
Wind Tolerance	★★★★
Salt Tolerance	★★
Drought Tolerance	★★★★
Flood Tolerance	★★
Cold Tolerance	★★

The Kei apple is a minor subtropical fruit of fair quality. The fruit is rarely eaten out of hand and its primary use is as an ingredient in jellies and preserves. The species forms a handsome shrub. The plant is tough and will tolerate drought, poor soil, and moderate cold. Therefore, the Kei apple may be a suitable addition to gardens in central and south Florida where conditions are unfavorable for growing other fruit trees.

Known Hazards

The plant has long spines capable of causing mechanical injury.

114

GEOGRAPHIC DISTRIBUTION The Kei apple originated in Africa. Its native range extends from the Cape of Good Hope north to Zimbabwe. It was introduced into Florida prior to 1900.

PLANT DESCRIPTION The Kei apple is a small evergreen tree or large shrub. The foliage is dense, deep green, and glossy. While the Kei apple reportedly grows to a height of 30 feet, most specimens in Florida are less than 10 feet tall. The plant can be trained as a low hedge. Leaves are oblong obovate, measuring between 1 1/2 and 3 inches in length. The Kei apple has a medium rate of growth. It is an attractive landscape plant with multiple uses. Because the Kei apple is armed with sharp thorns, it can form an impenetrable hedge. Small flowers with yellow stamens form in the leaf axils. Male and female flowers may be borne on separate plants, and only those plants bearing female flowers set fruit. Fruit set is enhanced if a male plant is planted in the vicinity of a female plant. However, isolated female plants have been observed to fruit heavily.

FRUIT CHARACTERISTICS The globose, yellow fruit measures between 1 and 2 inches in diameter. The skin is thin, smooth, and lightly pubescent. The flesh is aromatic and juicy but is also somewhat astringent. The fruit improves considerably during the later stages of ripening. About a dozen flat, pointed seeds are embedded in the flesh.

CULTIVARS No cultivars are regularly available in Florida. Most plants are grown from seed. At present, little interest exists in improving this species.

RELATIVES The Flacourtiaceae or Flacourtia family consists of about 90 genera and 1000 species. While the family contains many fruiting species, few produce fruit of great merit. Four species are grown in Florida with some frequency: the Abyssinian gooseberry, *Dovyalis abyssinica,* a shrub native to eastern Africa that bears a mildly astringent orange fruit; the ketembilla, *Dovyalis hebecarpa,* a vigorous shrub native to Sri Lanka that bears a purple fruit of fair quality; the *Dovyalis* hybrid or tropical apricot, *Dovyalis abyssinica x hebecarpa,* which bears acidic, velvety, orange-brown fruit; and the governor's plum, *Flacourtia ramontchi,* native to Africa and southern Asia. The governor's plum has been classified as an invasive exotic in Florida. Other fruiting species include the common sourberry, *Dovyalis rhamnoides,* of Africa; the Batoko plum, *Flacourtia inermis,* of southern Asia or the Philippines; the fried egg tree, *Oncoba spinosa,* of Africa; the kangu, *Flacourtia sapida,* of India; the paniala, *Flacourtia dataphracta,* of India; the rukam, *Flacourtia rukam,* of India and southeast Asia; and the wild apricot, *Dovyalis zeyheri,* of Africa.

CLIMATE The Kei apple is subtropical in habit, but is not strictly limited to south Florida. It has been grown in protected areas as far north as St. Augustine. It is fairly common in the Clearwater area on Florida's west coast. The plant has survived temperature drops to 19° F.

CULTIVATION The kei apple is a low-maintenance plant. It requires well-drained soil but, beyond that, is undemanding and will grow on sand, limestone soil, and alkaline soil. It is drought tolerant and has low water needs. It is somewhat salt tolerant. It will grow in full sun or partial shade. The plant has few serious pests or diseases in Florida. The fruit is moderately susceptible to attack by the Caribbean fruit fly. The Kei apple is usually grown from seed.

HARVEST AND USE The Kei apple is moderately productive and may bear 2 crops in Florida. The fruit is harvested when fully colored. It can be eaten fresh if fully ripe. Sprinkling the fruit with sugar neutralizes the acidity. The Kei apple makes a superb jelly and can also be used to flavor drinks.

The vicious spines of the Kei apple make the plant well suited for use as a barrier hedge.

Kiwifruit

SCIENTIFIC NAME: *Actinidia* spp.
FAMILY: Actinidiaceae
OTHER COMMON NAMES: Chinese Gooseberry,
    Kiwi (Spanish)

Fruiting Calendar

JAN	FEB	MAR	APR	MAY	JUN	JUL	AUG	SEP	OCT	NOV	DEC

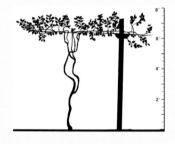

Characteristics

Overall Rating	★★
Ease of Care	★
Taste/Quality	★★★
Productivity	★★★
Landscape Value	★★
Wind Tolerance	★★
Salt Tolerance	★
Drought Tolerance	★★
Flood Tolerance	★★
Cold Tolerance	★★★★

In areas to which it is well suited, the kiwifruit is productive and well behaved. Unfortunately, the kiwifruit is not especially well suited to Florida. It is marginally productive in north Florida and only fruits sporadically in peninsular Florida. Hope remains, though, that better-adapted varieties will be found. While the kiwifruit may never assume the status of a commercial crop in Florida, it has the potential to become a popular and valuable dooryard crop.

Known Hazards

The fruit contains an allergen, which may cause contact dermatitis, oral allergies, and asthma, especially in those who have developed hypersensitivity from exposure to natural latex.

116

GEOGRAPHIC DISTRIBUTION The kiwifruit is native to eastern Asia. The fuzzy kiwifruit is native to southern China. Its relative, the hardy kiwifruit, originated in northern China, Manchuria, Siberia, and Korea. The kiwifruit rose from utter obscurity in the early 1900s to become an important fruit crop, enjoyed around the world by the close of the twentieth century. New Zealand first succeeded in commercializing the kiwifruit. Production within the United States is concentrated in California.

PLANT DESCRIPTION The kiwifruit is a deciduous woody vine or climbing shrub. Leaves are alternate and oval, usually 4 to 8 inches in length. The vine is supported by nonterminating shoots that coil around trellises or other objects. The vine is fast growing and vigorous, but may live as long as 50 years. Most varieties are dioecious, that is, they bear male and female flowers on separate plants. The yellow or cream-colored flowers, which form on new growth, are born in the leaf axils. They have 5 or 6 petals. Female flowers are typically somewhat larger than male flowers.

FRUIT CHARACTERISTICS The fruit of the kiwifruit is a berry, typically containing several hundred small seeds. The fruit of the common fuzzy kiwifruit is slightly smaller than a hen's egg. It has green skin coated with short bristles or hairs. Other species have thin edible skin that is hairless or that is coated with barely-perceptible down.

SPECIES AND CULTIVARS The fuzzy kiwifruit, *Actinidia deliciosa*, is the primary commercial species. Performance in Florida has been poor owing to the lack of winter chill. The cultivars 'Bruno,' 'Elmwood,' and 'Vincent' have fruited with some regularity in north Florida. 'Tomuri' is often used as a male pollinator. The smooth kiwifruit, *Actinidia chinensis*, may be suitable for planting in north and central Florida. It reportedly has low chilling requirements. Cultivars include 'California,' 'Hongyang,' 'Golden Yellow,' 'Lushanxiang,' and 'Tropical.' The hardy kiwifruit, *Actinidia arguta*, also has low chilling requirements and crops readily in the Florida panhandle. Promising cultivars include '74–8,' '74–49,' 'Anna,' 'Dumbarton Oaks,' 'Geneva,' 'Issai,' 'Ken's Red,' 'Lone Star,' 'Meador,' and 'Michigan.' The red kiwi, *Actinidia melanandra*, bears small, red, sweet fruit, and may have some potential as a crop for the subtropics.

RELATIVES The Actinidiaceae family consists of 4 genera and more than 200 species. The genus *Actinidia* contains about 70 species. At least 10 of these bear worthwhile fruit.

CLIMATE Lack of winter chill is a limiting factor to kiwifruit production in Florida. For reasons that are not fully understood, the kiwifruit seems to have greater chilling requirements in the southeastern United States than it does in California. At the same time, the vine may be damaged or killed when the temperature falls to 10° or 15° F.

CULTIVATION The kiwifruit is fairly demanding when it comes to maintenance. The vine prefers mildly acidic, well-drained loam. The optimal pH is 6.0, although the vines will make sufficient growth in soil with a pH of 5 to 6.5. Except when dormant, the plants require a steady supply of water. The kiwifruit requires periodic applications of high-nitrogen fertilizer from spring through mid-summer. It requires a trellis system or other support structure upon which to climb. The vine is susceptible to wind damage. Pruning is essential to fruit production. The idea is to encourage lateral growth along the trellis wires, with bearing canes shooting out toward the sides. Few serious diseases affect the kiwifruit. Minor pests include scale insects, thrips, and snails. The plant is most often reproduced through leafy or semi-hardwood cuttings, rooted under mist.

HARVEST AND USE In Florida, fruit ripen during the fall. The peel should be removed from the fuzzy kiwifruit. The hardy kiwifruit has edible skin and can be eaten whole. The kiwifruit has excellent storage characteristics. If refrigerated, it will keep for up to 2 months.

The hardy kiwi is smaller than the fuzzy kiwi, but has edible skin and superior flavor.

Kumquat

SCIENTIFIC NAME: *Fortunella* spp.
FAMILY: Rutaceae
OTHER COMMON NAME: Naranjita Japonés (Spanish)

Fruiting Calendar

JAN	FEB	MAR	APR	MAY	JUN	JUL	AUG	SEP	OCT	NOV	DEC

Characteristics

Overall Rating	★★★★
Ease of Care	★★★★
Taste/Quality	★★★★
Productivity	★★★★
Landscape Value	★★★★
Wind Tolerance	★★★★
Salt Tolerance	★★
Drought Tolerance	★★★
Flood Tolerance	★★★
Cold Tolerance	★★★★

The kumquat is a cold-hardy citrus relative, suitable for planting in all regions of Florida. The fruit is versatile and has good storage characteristics. Some varieties are tart and are primarily used as a flavoring agent or as an ingredient in marmalade. Other varieties are delicious eaten out of hand. The tree is compact and functions as an outstanding landscape specimen.

Known Hazards
Oils in the fruit rind and leaves may cause contact dermatitis or photosensitivity in some individuals.

GEOGRAPHIC DISTRIBUTION The kumquat is native to China and has been widely distributed to the world's subtropical and warm temperate regions. In the United States it is planted in California, Florida, south Georgia, and various Gulf states. A small commercial industry exists in Florida, with production centered in Pasco County.

TREE DESCRIPTION The kumquat is a compact evergreen tree, rarely exceeding 12 feet in height. The foliage is dark green and of medium texture. Leaves are alternate, measuring between 1 1/2 and 3 inches in length. They are toothed toward the apex.

FRUIT CHARACTERISTICS The fruit is small, measuring less than 2 inches in length. In form, it ranges from obovate to oval to globose. The rind, which is highly aromatic, adheres to the underlying flesh. The flesh is divided into several segments surrounding a solid core. Flavor ranges from sweet to mildly sour. Segments may contain several small seeds.

SPECIES AND CROSSES Four species are grown as fruit trees and ornamentals in the United States. Meiwa, *Fortunella crassifolia,* bears medium-size, rounded fruit of excellent flavor. It is the sweetest kumquat and is the best for eating out of hand. Nagami, *Fortunella margarita,* is frequently planted in north and central Florida. The fruit are oval, averaging about 1 1/2 inches in length. While edible out of hand, it does not equal Meiwa in flavor. Marumi, *Fortunella japonica,* bears a small round fruit, averaging just under an inch in diameter. The fruit may contain up to 3 small seeds. Hong Kong, *Fortunella hindsii,* is native to Hong Kong and surrounding regions of China. The tart fruit is tiny and round, averaging just over 1/2 inch in diameter.

Crosses between the kumquat and species from the *Citrus* genus have enjoyed varying degrees of success. 'Eustis' is a limequat—a hybrid between a kumquat and a lime. The flesh is yellow and relatively sweet. 'Nippon' is a cross between a kumquat and a mandarin. The fruit is moderately sweet and is of fair quality. 'Thomasville' is a citrangequat—a cross between a kumquat and citrange. The tree is a very cold hardy, and the fruit is of marginal quality.

RELATIVES The medium-size Rutaceae family includes all varieties of citrus and such diverse species as the bael fruit, *Aegle marmelos;* curry leaf tree, *Murraya koenigii;* elephant apple, *Feronia limonia;* limeberry, *Triphasia trifolia;* wampee, *Clausena lansium;* and white sapote, *Casimiroa edulis.* The kumquat and the mandarin exceed all other forms of cold-tolerant citrus in flavor. The trifoliate orange, *Poncirus trifoliate,* is the most cold-tolerant species of citrus. It is deciduous, very thorny, and produces acrid, inedible fruit. The calomondin, *Citrus madurensis,* resembles the kumquat in its habit of growth and outward appearance. However, the flesh is startlingly sour and is useful primarily in marmalades and preserves.

CLIMATE The kumquat has survived temperatures as low as 12° F. It is grown throughout all regions of Florida. Grafting on the rootstock of the trifoliate orange has been found to increase cold hardiness.

CULTIVATION The cultural requirements of the kumquat are similar to those of citrus. The tree is susceptible to attack by various insect pests, including mites and scales. Citrus leafminers sometimes disfigure the foliage. The kumquat does not appear to be particularly vulnerable to citrus canker. It can be reproduced by any method applicable to citrus.

HARVEST AND USE In Florida, the main crop ripens from October through early February. The fruit can be eaten out of hand and the peel is usually consumed along with the flesh. The kumquat makes excellent marmalade, chutney, or preserve, and is used to impart flavor to various recipes. It can also be skewered with meat and roasted over an open flame. The fruit has excellent keeping qualities and can be stored on the tree for up to 2 months without any diminishment in quality.

The 'Meiwa' kumquat bears a sweet fruit that is excellent for eating out of hand.

Lemon

SCIENTIFIC NAME: *Citrus limon*
FAMILY: Rutaceae
OTHER COMMON NAME: Limón (Spanish)

Fruiting Calendar

JAN	FEB	MAR	APR	MAY	JUN	JUL	AUG	SEP	OCT	NOV	DEC

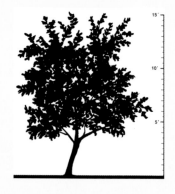

Characteristics

Overall Rating	★★★
Ease of Care	★★★
Taste/Quality	★★★
Productivity	★★★
Landscape Value	★★★★
Wind Tolerance	★★★
Salt Tolerance	★★
Drought Tolerance	★★★
Flood Tolerance	★★★
Cold Tolerance	★★★

The lemon is the world's most popular acid-citrus fruit. Several cultivars are reasonably well suited to Florida conditions. The tree is productive and bears fruit over much of the year. This quality, combined with the fact that the fruit has multiple uses, makes the lemon a good choice for dooryard planting in central and southern portions of the state.

Known Hazards

Oils in the fruit rind and leaves may cause dermatitis or photosensitivity in some individuals. Sharp spines are often present.

GEOGRAPHIC DISTRIBUTION The lemon is native to southern Asia, possibly India. It reached the Mediterranean prior to the year 1000 and was carried to the Americas by Christopher Columbus. During the late 1800s the lemon was an important commercial crop in Florida. However, commercial plantings have diminished. Today, the primary import of the lemon in Florida is as a dooryard crop.

TREE DESCRIPTION The lemon is a small, evergreen tree, usually attaining a height of 12 to 18 feet. Branches are often armed with sharp spines. The dark green leaves are alternate, elliptic in shape, and measure between 3 and 4 inches in length. Margins are toothed. The petioles are narrowly winged. Flowers measure about an inch in diameter, and have 4 or 5 white petals, often with a pink or purple blush on the underside.

FRUIT CHARACTERISTICS The fruit, which ranges from oval to elliptic in shape, measures between 2 1/2 and 5 inches in length. The mature fruit is usually yellow although some cultivars bear variegated fruit. The aromatic rind is of medium thickness. The flesh ranges from pale yellow to off-white, and is composed of between 8 and 10 segments, which may contain a scattering of seeds. The juicy flesh is tart and acidic, but is pleasant and delicious when sweetened or when combined with other flavors.

CULTIVARS Dozens of cultivars have been selected. Several are suitable for dooryard planting in Florida. 'Bearss' and 'Avon,' both selected in Florida, are excellent choices for the home garden. 'Meyer,' a cultivar discovered in China in 1908, is thought to be a cross between a lemon and mandarin. It performs well under Florida conditions. 'Ponderosa,' which may be a lemon x citron hybrid, produces a large, lumpy fruit and is occasionally grown in Florida. It is somewhat cold sensitive. 'Villafranco' was the leading commercial cultivar in Florida at one time. 'Eureka,' a California cultivar, and 'Lisbon,' a Portuguese cultivar, are often planted in Florida, but have not proven entirely satisfactory.

RELATIVES The family Rutaceae includes all varieties of *Citrus*. *Citrus* species discussed within this book include the grapefruit, *Citrus* x *paradisi;* the lime, *Citrus hystrix* and *Citrus aurantifolia;* the mandarin, *Citrus reticulata;* and the orange, *Citrus sinensis.*

CLIMATE The cold tolerance of the lemon is greater than that of the lime, but not equal to that of the orange. Fruit suffer damage at about 28°F. Defoliation and limb damage may occur with a temperature drop to 26°F. The 'Meyer' lemon appears to have greater cold tolerance than other varieties. High summer temperatures have an adverse impact on the lemon.

CULTIVATION General cultivation techniques appropriate for the orange can be applied to the lemon. Mites and scale insects frequently attack the tree, but can be readily controlled with approved pesticides. In Florida, the lemon is susceptible to several diseases, including anthracnose, branch knot, scab, greasy spot, gummosis, leaf spot, and various rots. Like other forms of citrus, the lemon can be propagated through budding or grafting.

HARVEST AND USE In Florida, the 'Meyer' lemon ripens from November through March. Other cultivars ripen over different periods. The lemon has good storage characteristics. If picked before maturity the fruit will store for several weeks with or without refrigeration. The fruit can be juiced or used as a garnish. The juice can be frozen and will store for many months. The pulp and juice serve as ingredients in numerous recipes, ranging from seafood dishes, to soups, to confectionaries and pies. The grated peel or zest serves as an aromatic spice. Many of the world's great cuisines make extensive use of the lemon.

The immature fruit of the 'Bearss' cultivar. This variety is well adapted to Florida growing conditions.

Lime

SCIENTIFIC NAME: *Citrus* spp.
FAMILY: Rutaceae
OTHER COMMON NAMES: Lima or Lima ácida (Spanish)

Fruiting Calendar

JAN	FEB	MAR	APR	MAY	JUN	JUL	AUG	SEP	OCT	NOV	DEC

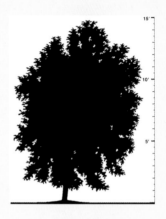

Characteristics

Overall Rating	★★★★
Ease of Care	★★★★
Taste/Quality	★★★
Productivity	★★★★
Landscape Value	★★★★
Wind Tolerance	★★★
Salt Tolerance	★★★
Drought Tolerance	★★★
Flood Tolerance	★★★
Cold Tolerance	★★

The common name "lime" applies to several *Citrus* species. These usually bear acidic fruit with skin that remains green through harvest. Two species—the Tahitian Lime and the Mexican Lime—flower and bear fruit throughout much of the year. The lime is highly recommended for planting as a dooryard tree in south Florida, where it has achieved popularity as a flavoring for drinks and as an ingredient in Key lime pie.

Known Hazards

Oils in the fruit rind and leaves have been known to cause severe dermatitis and photosensitivity. Sharp spines are often present.

122

GEOGRAPHIC DISTRIBUTION Like most members of the *Citrus* genus, the lime originated in Southeast Asia, most probably in the region encompassing the Malaysian peninsula and eastern portions of the Indian subcontinent. The Tahitian lime is grown commercially in Florida. The smaller and more aromatic Mexican lime or Key lime has faded as a commercial crop, but excels as a dooryard crop.

TREE DESCRIPTION The tree is evergreen, densely foliated, usually less than 15 feet in height. The branches of most varieties have prominent thorns. The dark green, leathery leaves are aromatic and are elliptic to oblong ovate in form. Flowers, which measure between 1 and 2 inches in diameter, have 4 to 6 petals. They are white, sometimes tinged with pink or purple. Some people are allergic to oils produced by the leaves, the rind of the fruit, and other parts of the plant. The reaction may be touched off by exposure to sunlight and may result in swelling, dermatitis, and other symptoms.

FRUIT CHARACTERISTICS The fruit is elliptic, usually measuring less than 3 inches in length. The skin typically turns from green to pale yellow at maturity. However, most fruit are picked before they attain physiological maturity. The pulp is contained in from 6 to 15 segments, which may contain a few to numerous seeds.

SPECIES The Mexican lime or Key lime, *Citrus aurantifolia*, forms a compact, shrublike tree. The fruit is small and seedy but is highly aromatic. The Tahitian lime or Persian lime, *Citrus latifolia*, is thought to be a hybrid of the Mexican lime and the lemon or citron. The sweet lime, *Citrus limettioides*, produces fruit that is nonacidic, but somewhat bland. It is rarely grown in Florida due to its susceptibility to viral diseases. The kaffir lime, *Citrus hystrix*, produces a highly acidic fruit with bumpy, yellow skin. The peel, juice, and leaves are used as ingredients in Thai cooking. The Rangpur lime, *Citrus x limonia*, bears a sour fruit of deep orange color.

RELATIVES Other members of the *Citrus* genus profiled within this book, include the grapefruit, *Citrus x paradisi*, the lemon, *Citrus limon*, the mandarin, *Citrus reticulata*, and the orange, *Citrus sinensis*.

CLIMATE As a general rule, the lime is not tolerant of freezing temperatures and is more sensitive to cold than the sweet orange, the grapefruit, and the lemon. The Tahitian lime is slightly hardier than the Mexican lime, but will suffer serious damage when the temperature falls to about 27° F. The Rangpur lime is cold hardy and will grow in north Florida.

CULTIVATION The lime is a low-maintenance tree. It grows well on sand and oolitic limestone, although it performs poorly on heavy clay. The Mexican lime is highly tolerant of drought and can withstand minor salt spray. The Tahitian lime is less drought tolerant. Red citrus mites, rust mites, broad mites, and scale insects occasionally bother the tree. Withertip, a fungal disease, affects the Mexican lime, but not the Tahitian lime. Scab sometimes develops during periods of wet weather. The tree is also vulnerable to citrus canker. Budding is often used as a propagation technique. However, the Mexican lime is often grown from seed, as seedlings show little variation.

HARVEST AND USE In south Florida, production occurs over large portions of the year. A mature tree may produce from 50 to 90 pounds of fruit. Fruit ripen about 6 months after flowering. The Mexican lime undergoes periods of high productivity in early summer and early winter. The Tahitian lime bears primarily during the summer and early fall. The Rangpur lime is productive over the winter. The fruit of the Mexican and Tahitian lime is usually harvested at the first color shift—from dark green to light green. The lime can be juiced and is used as a flavoring, marinade, garnish, or ingredient in a wide range of dishes.

The deep orange Rangpur lime will survive colder temperatures than most other acid citrus species.

Longan

SCIENTIFIC NAME: *Dimocarpus longan*
FAMILY: Sapindaceae
OTHER COMMON NAME: Dragon's eye

Fruiting Calendar

JAN	FEB	MAR	APR	MAY	JUN	JUL	AUG	SEP	OCT	NOV	DEC

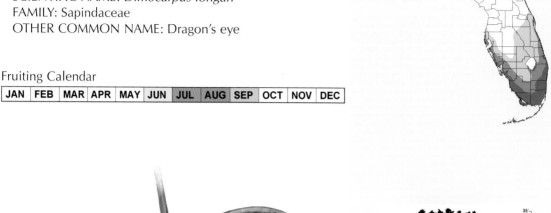

Characteristics

Overall Rating	★★★★★
Ease of Care	★★★
Taste/Quality	★★★★★
Productivity	★★★
Landscape Value	★★★
Wind Tolerance	★★★
Salt Tolerance	★
Drought Tolerance	★★★★
Flood Tolerance	★★
Cold Tolerance	★★★

The longan is a premium fruit and is one of the most rewarding species to grow in south Florida. When discussing the longan, it is difficult to avoid comparisons with its glamorous cousin, the lychee. Some authorities have characterized the longan as inferior to the lychee. However, the flavor of both species is excellent. The tree of the longan is more forgiving of error than that of the lychee. It is hardy, drought tolerant, and productive.

Known Hazards

None

GEOGRAPHIC DISTRIBUTION The longan originated in Southeast Asia, possibly in southern China or Myanmar. It was first introduced into Florida in 1903, but aroused little interest in the first half of the twentieth century. In 1954 rare-fruit pioneer William Whitman introduced the superior 'Kohala' cultivar from Hawaii. Since then, appreciation for this species has been growing. Commercial production is presently concentrated in Miami-Dade County.

TREE DESCRIPTION The longan is a medium evergreen tree. It typically attains a height of between 30 and 45 feet. The canopy is bushy, dense, and rounded. The habit of growth ranges from erect to spreading. Leaves are alternate, compound, and pinnate. They are composed of between 7 and 15 leaflets. The leaflets are glossy, dark green above, and pale green below. They are ovate-oblong and measure between 4 and 10 inches in length. Emerging foliage may be scarlet. The bark is grayish-brown, corky, and rough. The rate of growth is moderate. The longan makes a handsome landscape specimen.

Flowers are born on upright, branched panicles. These emerge from the terminal buds of recent growth. In Florida, bloom typically occurs from late February through April. Some growers have induced off-season flowering by applying low concentrations of potassium chlorate as a foliar spray. Flowers are small and brownish-yellow in color. Each has 2 calyx lobes and 5 petals. Flowers may be male, female, or hermaphrodite, and various types are borne on a single panicle. Flowers are produced sequentially: the male flowers form first, followed by the female flowers, followed by the hermaphrodite flowers, followed by another wave of male flowers. The tree has some tendency toward biennial bearing, although this defect is less pronounced than in the lychee.

Numerous insects visit the flower panicles of the longan.

FRUIT CHARACTERISTICS The fruit is a globose drupe, measuring between 3/4 and 1 1/2 inches in diameter. The rind is thin, leathery, and lightly textured. As the fruit matures, the rind becomes somewhat brittle. The color changes from green to light brown or yellowish-tan. The aril is fleshy and translucent. It surrounds a single dark brown, ovoid seed. In some cultivars, a high percentage of fruit contains small or aborted seeds. Superior cultivars have crisp, thick flesh adhering to neither the seed nor the rind. The flavor is sweet, musky, and mildly aromatic.

The fruit of the 'Kohala' cultivar has a small seed and thick firm flesh.

CULTIVARS More than 400 cultivars have been selected in Southeast Asia and elsewhere. On the whole, cultivars from Thailand are considered superior to those from other regions—perhaps owing to that nation's advanced breeding program and general enthusiasm toward the fruit. Several cultivars consistently bear high quality fruit in Florida.

'Kohala' - This Hawaiian selection is the primary commercial and home-garden cultivar in Florida. It offers consistent production and large fruit of sweet flavor and good quality. Most agree that 'Kohala' remains the best all-around commercial cultivar. The fruit matures early, usually in July.

'Biew Kiew' - This Thai variety bears large fruit of superior quality. Indeed, the flavor is widely regarded to be among the best of any of the longans. Thus far, its bearing habits have been steady in Florida. It is a mid-season variety, with fruit ripening from late July through August.

'See Chompoo' - This Thai variety exceeds Kohala in overall quality, but its bearing habits have not been adequately assessed. The aril is reportedly pink-tinged in Thailand. In Florida, this trait has not been evident. The fruit is firm fleshed, has a small seed, and has an excellent, perfumed flavor. Fruit ripen from late July through August.

'Tiger's Eye' - This cultivar produces a high quality fruit similar in many respects to that produced by 'Kohala.' It is a remarkably heavy bearer in some years. 'Tigers Eye' is a mid-season longan.

'Edo' - This cultivar, also called 'Daw,' is a leading Thai commercial cultivar. The fruit is sweet, medium to medium-large and of good flavor. The tree is a prolific bearer and may eventually be adopted as a commercial cultivar in Florida. This is a mid- to late-season variety, with fruit ripening in August and early September.

'Diamond River' - This Australian cultivar produces a medium-size fruit of excellent flavor. However, the surface of the aril is somewhat slick and watery. It is a prolific producer. This cultivar is reported to set multiple crops within a single year in some locations. 'Diamond River' has also shown some ability to withstand flooding. It is a late-season cultivar, ripening in September.

A Vietnamese variety, 'White Pepper Seed,' has recently been introduced into Florida. It bears a small fruit with a light rind and a tiny seed. Other cultivars that may hold promise for planting in Florida include 'Baidum,' 'Chuliang,' 'Dagelman,' 'Dang,' 'Egami,' and 'Xuong com Vang.'

RELATIVES The Sapindaceae or Soapberry family, within the order Sapindales and the subclass Rosidae, consists of about 125 genera and 1,000 species. Well-regarded members include the lychee, *Litchi chinensis,* profiled in a subsequent section; the rambutan, *Nephelium lappaceum;* and the pulasan, *Nephelium mutabile.* The rambutan and pulasan are adapted to a more tropical climate than that found in peninsular Florida. The alupag, *Dimocarpus didyma* or *Dimocarpus longan* spp. *philippensis,* is native to the Philippines. It produces a sweet, rough-skinned fruit.

The pulasan, a longan relative, is tropical in habit and is too cold tender to grow on the Florida peninsula.

The longan is distantly related to the akee, *Blighia sapida,* originally from West Africa. The akee is considered the national fruit of Jamaica. It is well suited to growth in south Florida. However, the fruit itself is poisonous until it has been exposed to sunlight. It is therefore critical that the capsule—which contains the edible arils surrounding the seeds—split open or "yawn" prior to harvest. The seeds, capsule, membranes, and other portions of the plant are toxic and must be discarded. When harvested and prepared under strict safety precautions, the akee can make a delicious addition to the table.

The mamoncillo, *Melicocca bijuga,* also known as the Spanish lime or genip, is native to South America. The green-skinned fruit measures about an inch in diameter. It is borne in large hanging clusters. The fruit is aromatic and pleasantly tart, although the flavor is inferior to that of the longan. The flesh is scant and surrounds a large seed. The cultivar 'Queen,' which originated in Key West, is self-compatible and produces a fruit of superior size and quality.

CLIMATE The longan is subtropical in habit. An established tree may survive a temperature drop to about 25° F, with defoliation and extensive twig damage. Experience in Florida has failed to confirm reports that it has greater cold tolerance than the lychee. The longan is grown as far north as Merritt Island on Florida's east coast and Sarasota on Florida's west coast.

CULTIVATION The longan thrives on well-drained sandy loam, but is not particular as to soil type. It will grow on oolitic limestone, where it appears to be less prone to iron deficiency than the lychee. The tree is intolerant of salt in the soil or water. It cannot withstand prolonged flooding and makes poor growth in heavy clay. The tree is moderately drought tolerant. Regular irrigation should be pro-

The stately 'Kohala' longan provides dense shade.

vided from the emergence of blooms in the spring until the advent of the rainy season.

Mulch should be applied in a thick layer beneath a young tree. Competing weeds and grasses should be eliminated from beneath the canopy. Fertilizer should be withheld from newly planted trees for at least 2 months. A mature tree should receive periodic applications of balanced fertilizer throughout the production cycle. The final application, which usually takes place around the time of harvest, may contain extra nitrogen to stimulate a post-harvest growth flush.

The tree can be kept to a manageable size through judicious pruning. Harvesting is really a form of light pruning, since it involves removal of the branch tip. By severing the branch further back along its length than would normally be required, the grower can harvest and prune at the same time. A fruit cluster should not contain more than 75 to 100 fruit. Where panicles contain excess fruit, the grower should sever the tip of each panicle or should carefully thin immature fruit by hand, leaving 1 or 2 fruit on each branchlet. This task is conducted when fruit reach the size of a pea. Thinning excess fruit increases fruit size and reduces tendencies toward biennial bearing.

PESTS AND DISEASES The lychee webworm causes significant damage to the longan in Florida. It is thought that this moth was introduced into Florida from the Caribbean. The moth lays eggs adjacent to newly formed buds. Upon hatching, the worms bore into new tissue, destroying terminal growth and flower buds. The symptoms are a sudden dieback or wilting of new growth and fine webs covering emerging leaves and flowers. The grower should conduct periodic inspections, looking for webs, eggs, or evidence of dieback. Several pesticides are registered for use on the longan in Florida. Other significant insect pests include scale insects and root weevils.

Growers inevitably suffer heavy losses from birds and squirrels. The boat-tailed grackle has a particular affinity for the fruit. Exclusion netting may be an acceptable solution for a small tree. Scare devices are somewhat effective. As the tree matures, production will gradually outstrip the damage inflicted by most pests.

PROPAGATION In Florida, air layering is by far the most popular method of propagation. It is usually carried out over the summer. The air layers are severed from the parent tree after about 2 months. The new tree is then placed in partial shade for between 4 and 6 months before it is moved into full sun. Air layers taken from the cultivar 'Kohala' often develop a vertical seam in the trunk just above ground level. This minor defect disappears as the tree matures. The longan can also be reproduced through budding, cuttings, and grafting. While the tree will grow readily from seed, seedlings are slow to come into bearing and often produce inferior fruit with scant flesh.

HARVEST AND USE A mature tree may yield between 100 and 400 pounds of fruit. The fruit matures 5 or 6 months after flowering. Harvest generally occurs from mid-summer to early fall. If picked too soon, the fruit will be flavorless and will not improve. If the fruit remains on the tree past maturity, it will develop an off flavor and the aril will take on a yellowish cast. The panicles are cut from the tree with pruning sheers or a V-shaped blade on a pole. Clusters should not be permitted to fall to the ground. The fruit has a short shelf life, and the rind will begin to harden and turn dark brown within 2 days after harvest. Refrigerated fruit will store for about a week. The fruit can be frozen, canned, or dried. The dried fruit is delicious and very sweet. It can be dried within its rind, to form a "longan nut" or without the rind, to form a "longan raisin."

A cluster of sweet 'See Champoo' longans is ready to harvest.

Loquat

SCIENTIFIC NAME: *Eriobotrya japonica*
FAMILY: Rosaceae
OTHER COMMON NAMES: Japanese Plum, Nispero
  Japonés (Spanish)

Fruiting Calendar

JAN	FEB	MAR	APR	MAY	JUN	JUL	AUG	SEP	OCT	NOV	DEC

Characteristics

Overall Rating	★★★
Ease of Care	★★★★
Taste/Quality	★★★
Productivity	★★★
Landscape Value	★★★★
Wind Tolerance	★★★
Salt Tolerance	★★
Drought Tolerance	★★★
Flood Tolerance	★★★
Cold Tolerance	★★★★

The loquat is a handsome tree that produces a delicious fruit. The fruit resembles a dessert plum in taste. Unfortunately, the loquat is a prime host of the despised Caribbean fruit fly. In many areas the fruit is rarely fit for human consumption. Where the fly is numerous, the loquat cannot be considered as anything beyond an attractive landscape specimen. However, in northern parts of the peninsula, the loquat remains an important door-yard fruit.

Known Hazards

The seeds are said to contain toxins and the leaves contain traces of arsenic.

GEOGRAPHIC DISTRIBUTION The loquat originated in China and has been grown in Japan for at least 1,000 years. It is a common fixture in Florida's suburban landscape.

TREE DESCRIPTION The loquat is a small evergreen tree with a dense, rounded crown. It can attain a height of about 30 feet. The foliage is ornamental, course, and deeply textured. Leaves are lanceolate, with serrate margins. They typically measure between 6 and 12 inches in length. The tree has a short to medium lifespan and is productive for about 30 years. Flowers are white or off-white, fragrant, and moderately showy. Most have 5 petals, 5 pistils, and 20 stamens. Some cultivars are self-fertile while others require a nearby pollinator.

FRUIT CHARACTERISTICS The fruit is a pome, measuring 1 to 2 inches in length. It may be pyriform or ovoid. It forms loose clusters, usually containing between 4 and 20 fruit. The skin is smooth and pubescent, usually yellow or orange in color. The flesh is soft, moist, either yellow or white, and contains between 1 and 5 medium brown seeds. The flavor is sweet, sometimes slightly tangy, but always mild and pleasant.

CULTIVARS Cultivars suitable for dooryard planting in Florida include 'Wolfe,' a yellow-fleshed, early-season cultivar, that bears large fruit with a low seed count; 'Tanaka,' an orange-fleshed, late-season cultivar; 'Golden Nugget,' an orange-fleshed, late-season cultivar; 'Premier,' a late-season cultivar with yellow skin and white flesh; 'Oliver,' a productive cultivar with some cold tolerance; 'Christmas,' a very early cultivar that produces high-quality fruit; 'Champagne,' a yellow-skinned cultivar that produces heavier crops when cross fertilized; 'Advance,' a mid-season cultivar that requires cross-pollination; 'Bradenton,' a quality mid-season cultivar; and 'Vista White,' a well-regarded late season cultivar.

RELATIVES The loquat belongs to the large and diverse Rosaceae family. Besides the loquat, this book features 8 other members of the Rosaceae family: apple, *Malus domestica;* blackberry, *Rubus* spp.; capulin, *Prunus salicifolia;* chickasaw plum, *Prunus angustifolia;* mayhaw, *Crataegus* spp.; peach, *Prunus persica;* pear, *Pyrus* spp.; and strawberry, *Fragaria ananassa.*

CLIMATE The loquat requires a subtropical to mild-temperate climate. Established trees in the Tallahassee area have survived low temperatures of 10° F. However, because flowering and fruit set occurs over the winter, freezing temperatures may limit fruit production. The buds, flowers, and fruit are susceptible to injury between 22° and 26° F. The loquat will grow and fruit as far south as the Keys.

CULTIVATION The loquat is an extremely low-maintenance tree. It will grow on a wide variety of soils, but will not tolerate poor drainage. The tree is drought tolerant, but will produce higher quality fruit if it receives regular irrigation. The loquat is somewhat wind tolerant, but has a tendency to blow over in hurricane-force winds. The most serious pest in south Florida is the Caribbean fruit fly. Cultivars that mature early, while the weather is still cold, are less likely to suffer infestations. Birds are attracted to the fruit and are responsible for significant losses. Fire blight is the most important disease affecting the plant. Infected branches should be severed well below diseased portions and removed from the site. Pruning sheers should be decontaminated in a bleach solution after each cut. The loquat is usually reproduced through grafting. Side veneer grafts and cleft grafts are commonly used.

HARVEST AND USE Typical yield is between 50 and 150 pounds per tree. In Florida, harvest ranges from February through May, depending on the cultivar. Entire panicles are clipped from the branch. The fruit is fragile and easily bruised, so must be handled with care. It will store in the refrigerator for about 10 days. The loquat may be eaten out of hand. It can also be made into pies or preserves.

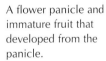

A flower panicle and immature fruit that developed from the panicle.

Lychee

SCIENTIFIC NAME: *Litchi chinensis*
FAMILY: Sapindaceae
OTHER COMMON NAMES: Mamoncillo Chino or Lechia
(Spanish)

Fruiting Calendar

JAN	FEB	MAR	APR	MAY	JUN	JUL	AUG	SEP	OCT	NOV	DEC

Characteristics

Overall Rating	★★★★★
Ease of Care	★★
Taste/Quality	★★★★★
Productivity	★★
Landscape Value	★★★
Wind Tolerance	★★★
Salt Tolerance	★
Drought Tolerance	★★★
Flood Tolerance	★★★
Cold Tolerance	★★★

China's most-revered fruit, the lychee, is reasonably well adapted to growing conditions in south Florida. The fruit itself is delicious. It provides the perfect balance of sweetness and acidity in a mouthful of luscious pulp. The tree, though, has several shortcomings. It is notorious for irregular yields and is fussy in its cultural requirements. Nevertheless, with its showy fruit and deep green foliage, the lychee is an outstanding choice for dooryard planting in south Florida.

Known Hazards

None

GEOGRAPHIC DISTRIBUTION The lychee originated in southern China, where it has been cultivated for thousands of years. While China remains the largest producer, the lychee is now grown throughout the world's subtropical regions. Centers of production include India, South Africa, and Australia. The worldwide market for fresh fruit has expanded rapidly over the last decade and demand has outstripped production.

Lychee seedlings were introduced to Florida from India in the early 1880s. However, the species languished in obscurity until the 1940s. Soldiers and civilian workers who served in the Pacific Theater during World War II returned home with fond memories of this fruit. The advent of air layering as a method of propagation and the introduction of the 'Brewster' cultivar sparked commercial interest. During the late 1940s and early 1950s, extensive orchards were planted around Sarasota. A severe freeze in 1958 nearly wiped out the industry. Lychee production moved south and is now concentrated in Miami-Dade County. Small commercial plantings are also located on Pine Island, Merritt Island, and in western Broward and Palm Beach Counties.

TREE DESCRIPTION The lychee is a medium to large evergreen tree. Depending on the cultivar, it may attain a height of between 25 and 70 feet. The crown is densely foliated, roughly symmetrical, round, and spreading. Leaves are alternate, pinnately compound, made up of 4 to 8 leaflets. Leaflets are lanceolate, pointed at the tips. They are deep green and glossy above, lighter below, of medium texture. Leaflets measure between 2 and 3 1/2 inches in length. While a seedling tree has deep,

The 'Emperor' lychee produces small flower panicles, which are followed by very large fruit.

vertical taproots, the roots of an air-layered tree are concentrated near the surface of the soil. Flowering takes place in February and March. Flowers are small and greenish yellow. They are borne profusely on many branched panicles that emerge from the terminal buds.

Most cultivars fruit irregularly from year to year. Cool winter temperatures tend to induce flowering. The amount of time between the most recent flush of growth and the onset of flowering also appears to have a significant impact on production. A tree that has not put out new growth for 3 months prior to flowering will usually produce a heavy crop. A tree that puts out new vegetative growth during the winter will not produce a heavy crop during the spring.

FRUIT CHARACTERISTICS The heart-shaped or oval fruit measures between 1 and 2 inches in length. The skin is leathery and rough in texture, covered with tiny tubercles or raised plates. Color ranges from red, to pink, to pinkish-green. The skin peels easily to reveal a whitish-translucent aril, which surrounds a single seed. The pulp is juicy, somewhat grapelike, but soft and melting. The flavor is sweet, subacid, aromatic, and exquisite. The seed is loosely embedded within the flesh. It is hard and oblong with a glossy, dark brown coat. The seed ranges from 1/2 to 1 inch in length. Varieties with smaller seeds are favored, since the flesh recovery ratio is higher. Some varieties have a large percentage of shrunken "chicken tongue" seeds.

Emerging foliage of the 'Mauritius' cultivar.

CULTIVARS Hundreds of varieties have been selected in China and other parts of the world. Of these, about 20 have achieved commercial import. The following cultivars have shown themselves to be well suited to dooryard production in Florida.

'Sweet Heart' - This cultivar, selected in Florida, is a consistent producer of quality fruit. The deep red fruit is medium large. The flavor is very good and fruit have a very high percentage of aborted seeds. The habit of growth is somewhat sprawling. Fruit typically ripen at the end of June.

'Mauritius' (also 'Kwai Mi' or 'Kuei Wei') - 'Mauritius' is the number-one commercial cultivar in Florida and is consistently productive. The light-red fruit is medium-large. Flavor is very good to excellent. 'Mauritius' ripens from the last week of May through the first 2 weeks of June in Florida. The tree is susceptible to damage from hurricane winds. The fruit is susceptible to damage by anthracnose.

'Emperor' - This cultivar produces an exceptionally large, red, heart-shaped fruit of good flavor. Seeds tend to be large, although the fruit still provides ample flesh. 'Emperor' has shown itself to be a reasonably consistent bearer in Florida. The fruit typically ripens in the third or fourth week of June. The small, slow-growing tree has a rounded, dense crown of medium-green foliage.

'Hak Ip' - 'Hak Ip' or 'Black Leaf' has been somewhat inconsistent in its bearing habit in Florida. However, because of the superb quality of the fruit, it is recommended as a cultivar for the home garden. The red fruit is of medium size.

A cluster of immature 'Hak Ip' lychees.

'Bosworth 3' - (also 'Kwai May Pink')–'Bosworth 3,' a top commercial cultivar in Australia, has shown itself to be a consistent bearer in Florida. It produces small to medium, pink-skinned fruit of superior flavor. Fruit ripen during the second half of June.

'Brewster' (also 'Chen Zi', 'Chen Tzu') - Brewster bears bright red, wedge-shaped fruit of medium size and good flavor. The tree is large, vigorous, and upright. Leaf tips tend to turn brown during dry spells, but this tendency does not impair yield. 'Brewster' is somewhat irregular in its bearing habit, tending to produce a heavy crop every second or third year. Fruit ripen in the second and third weeks of June.

'Sweet Cliff' (also 'T'im ngam' or 'T'ien yeh') - 'Sweet Cliff' bears small pink fruit of good flavor with a slight musky quality. This cultivar struggles on the shallow limestone soil of Miami-Dade County. However, it is a very consistent producer in other parts of south Florida. The fruit ripen from the last week of May through the first week of June.

Other cultivars which display desirable characteristics include 'Ambiona,' 'Bengal,' 'Early Large Red,' 'Groff,' 'Ohia,' and 'Yellow Red.'

RELATIVES The lychee is a member of the Sapindaceae family. The only other member included within this book is the longan, *Dimocarpus longan*, profiled earlier. Other members of the Sapindaceae family are described in the subsection pertaining to relatives of the longan.

CLIMATE The cold tolerance of the lychee is similar to that of the longan. A mature tree will sustain damage at about 28° F, but may survive temperatures as low as 25° F. The lychee grows as far north as Merritt Island, on Florida's east coast, and Sarasota, on Florida's west coast. It is also grown in Highlands County and other parts of the interior of south central Florida.

CULTIVATION The lychee has intermediate maintenance requirements. The tree is touchy when young, but becomes quite hardy with age. It will succeed on a wide range of soils, although it prefers a soil with a pH of 6 to 7. It is intolerant of salt in the soil, air, or water.

The lychee requires adequate drainage, but can tolerate brief flooding. The tree benefits from periodic irrigation during establishment, during times of drought, and during periods of fruit development. In Florida, the summer rainy season is usually sufficient to ensure an adequate post-harvest flush of vegetative growth.

The lychee is moderately wind resistant. While Hurricane Andrew's category 5 winds caused significant damage to lychee trees in the Homestead area, a surprising percentage survived. Losses were most severe among trees of the 'Mauritius' cultivar, which has a weak branch structure. As long as the root system remains intact, the tree has a remarkable ability to recover from wind-related damage.

When young, the tree is susceptible to damage from over fertilization. Inexperienced growers will often kill a young tree by attempting to push it along. No fertilizer should be applied for at least 2 or 3 months after the tree has been planted. Thereafter, the grower may begin light applications of a slow-release fertilizer. The lychee sometimes becomes chlorotic on the oolitic limestone of southeastern Florida and may require applications of chelated iron. Fertilization is suspended from October to February to discourage vegetative growth over the cooler months. This has been found to promote flowering and fruiting.

Girdling has been found to increase fruit set over the near term. This technique requires the removal of a narrow ring of bark, no more than 1/4 of an inch wide, from 1 or 2 major branches. Only part of the tree should be girdled at any one time. Unless a bark connection remains between parts of the canopy and the ground, the tree will die. Girdling is most effective if conducted in September or early October.

PESTS AND DISEASES The lychee has a few serious insect pests in Florida, although most can be controlled with approved pesticides. The lychee webworm, described in connection with the longan, is a problem in some areas. This insect appears to be most active from November through February. Root weevils cause considerable damage. Larvae feed on the roots, often leading to a decline in tree vigor. The mature insects cut rounded notches in leaf margins. Other minor pests include various scale insects, stinkbugs, larva of the cotton square borer, and aphids. Birds and squirrels cause significant losses. Scare devices, such as owl replicas, noisemakers, dangling compact disks, Mylar balloons, and metallic ribbons have been used with some success. However, every grower should be prepared to lose fruit to these persistent raiders. Anthracnose, a fungal disease, sometimes affects maturing fruit and can cause skin discoloration and off flavor. Oak root rot can kill a tree planted on land formerly occupied by oaks.

PROPAGATION In Florida, the lychee is reproduced almost exclusively by air layering. Seedlings are variable and rarely approach the quality of the cultivar from which the seeds were taken. The best branches for air layering are those at the periphery of the canopy. After it is removed from the parent tree, the air layer is kept in 70 percent shade for 2 months before it is gradually exposed to more intense light.

HARVEST AND USE In Florida, the lychee ripens in May and June. Yield from a 20-year-old tree averages between 50 and 150 pounds. The fruit must be fully ripe when harvested, but will deteriorate if it is left on the tree for more than a week once it has attained full coloration. The entire panicle is clipped from the branch. The fresh fruit has a short shelf life. At room temperature, it remains attractive for no more than 2 or 3 days. The fruit can be frozen for several weeks, but it tends to deteriorate rapidly once it has been thawed.

These delicious 'Hak Ip' lychees are almost ready to harvest.

Mabolo

SCIENTIFIC NAME: *Diospyros blancoi*
FAMILY: Ebenaceae
OTHER COMMON NAME: Velvet Apple

Fruiting Calendar

JAN	FEB	MAR	APR	MAY	JUN	JUL	AUG	SEP	OCT	NOV	DEC

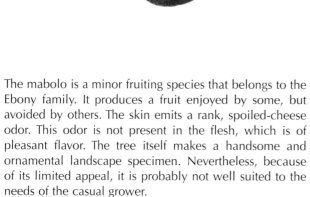

Characteristics

Overall Rating	★★
Ease of Care	★★★
Taste/Quality	★★★
Productivity	★★
Landscape Value	★★★
Wind Tolerance	★★★
Salt Tolerance	★
Drought Tolerance	★★★
Flood Tolerance	★★★
Cold Tolerance	★

The mabolo is a minor fruiting species that belongs to the Ebony family. It produces a fruit enjoyed by some, but avoided by others. The skin emits a rank, spoiled-cheese odor. This odor is not present in the flesh, which is of pleasant flavor. The tree itself makes a handsome and ornamental landscape specimen. Nevertheless, because of its limited appeal, it is probably not well suited to the needs of the casual grower.

Known Hazards

None

GEOGRAPHIC DISTRIBUTION The species is native to the Philippines, where it is well regarded and frequently planted. It has spread to Southeast Asia and other tropical areas, although it has not been widely embraced as a fruit tree in any location outside its homeland. It was first introduced into Florida in 1906, but is rarely seen outside of rare fruit collections.

TREE DESCRIPTION The mabolo is typically a small to medium evergreen tree, attaining a height of 30 to 40 feet. It can reach a larger size in its tropical homeland. The oblong leaves are alternate, thick, and glossy. They measure between 5 and 9 inches in length. The trunk is straight; the bark is dark and furrowed. The rate of growth is moderate. White, tubular flowers appear in the spring. The mabolo is dioecious; that is, male and female flowers are born on separate trees. The presence of a male pollinator is required to ensure adequate fruit set.

FRUIT CHARACTERISTICS The unusual and attractive fruit is oblate, measuring 2 to 3 1/2 inches in diameter. A 4-lobed calyx is present at the base. The skin is burnt orange or purple-red in color, thickly coated with short, glistening, coppery bristles. The skin, though relatively thin, is peeled prior to consumption. The flesh is white, crisp, and moist. The flavor is mild, sweet, and pleasant, perhaps resembling a firm-fleshed banana with a hint of cheese. Although the fruit is sometimes seedless, it may contain as many as 8 large seeds radiating from the central core.

CULTIVARS Few cloned varieties are available in Florida. The cultivars 'Manilla' and 'Valesca' bear quality fruit that is usually seedless.

RELATIVES The mabolo belongs the Ebonaceae or Ebony family. Within this family are the black sapote, *Diospyros dignia,* and persimmon, *Diospyros kaki,* both profiled within this book. The American persimmon, *Diospyros virginiana,* is described in the subsection pertaining to relatives of the persimmon. The jackal berry, *Diospyros mespiliformis,* an African species, bears sweet fruit about an inch in diameter. The date plum, *Diospyros lotus,* is a deciduous, temperate tree from Asia. It bears small yellow fruit, tart in flavor.

CLIMATE The cold tolerance of the Mabolo has not been adequately assessed; however, its tropical origins suggest that it would succumb to winter cold in central parts of the peninsula. It suffers significant foliage damage when the temperature falls to 28° F.

CULTIVATION The mabolo requires very little in the way of cultural attention. It is drought tolerant, requiring irrigation only during the establishment phase. It will readily adapt to various soil types. It produces adequate crops without fertilization. The tree is typically grown from seed, although shield budding may be used to perpetuate superior varieties.

HARVEST AND USE In Florida, the mabolo ripens in the summer and early fall. Maturity is indicated when the fruit reaches full size and coloration. The fruit is usually clipped from the tree. It is probably at its best when peeled, chilled, and quartered.

The flower and developing fruit of the mabolo.

Macadamia

SCIENTIFIC NAME: *Macadamia* spp.
FAMILY: Proteaceae
OTHER COMMON NAMES: Queensland Nut, Australia Nut

Fruiting Calendar

JAN	FEB	MAR	APR	MAY	JUN	JUL	AUG	SEP	OCT	NOV	DEC

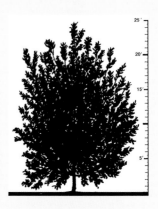

Characteristics

Overall Rating	★★
Ease of Care	★★★
Taste/Quality	★★★★
Productivity	★
Landscape Value	★★★
Wind Tolerance	★★★
Salt Tolerance	★
Drought Tolerance	★★★
Flood Tolerance	★★★
Cold Tolerance	★★

The macadamia is one of the world's premier nut species. It has some potential as a dooryard crop in Florida. It grows vigorously, experiences few problems, and makes a handsome landscape specimen. Unfortunately, production is sparse. A large tree will produce only about 25 pounds of nuts each year. The average tree in Hawaii produces more than 100 pounds of nuts. Perhaps, in the future, cultivars capable of higher yields will be selected for Florida.

Known Hazards

None

GEOGRAPHIC DISTRIBUTION Two species produce an edible nut. Both are native to the east coast of Australia. The macadamia was first introduced into Hawaii in the late 1870s or early 1880s. Perfect climatic conditions and the development of superior clones soon established Hawaii as the hub of worldwide production and distribution. No commercial production of any consequence takes place in Florida.

TREE DESCRIPTION The macadamia is a medium-size evergreen tree with good ornamental value. In Florida, it rarely exceeds 35 feet. Leaves are dark green, leathery, narrow, and measure from 8 to 14 inches in length. The trunk is short and the bark is rough and grayish-brown. Flowers are borne on pendulous racemes, which emerge from leaf axils. These may be from 5 to 15 inches in length. The small tubular flowers may be white, yellow, or pink. Most trees are partially self-sterile and benefit from cross-pollination.

The 'Beaumont' macadamia flowers profusely. The pink flowers show that it has some rough-shelled macadamia heritage.

FRUIT CHARACTERISTICS Each stalk may support up to 25 nuts. A green husk encloses the round, deep brown nut. The husk splits open as the nut matures. The nut has an extremely hard shell. The kernel is creamy white and oval. It may be round or may be divided into hemispherical kernels. The meat is delicious, moderately firm, sweet, and crunchy.

SPECIES AND CULTIVARS The smooth-shelled macadamia, *Macadamia integrifolia*, bears a nut with high oil content, suitable for commercial use. Flowers are white. Leaves, arrayed in whorls of 3, are entire. The rough-shelled macadamia, *Macadamia tetraphylla*, bears a nut with a pebbled shell, high carbohydrate content, and low oil content. Leaves, arrayed in whorls of 4, have serrated margins. Flowers are usually pink or yellowish-white.

More than 50 cultivars have been selected. In Florida, high yield is the most important considera-

tion. Four cultivars are recommended: 'Dana White,' a productive cultivar selected in Homestead; 'Beaumont,' a hybrid selected in Australia with spiny leaves, pink flowers, and reasonable productivity; 'Arkin Papershell,' a productive cultivar that produces a high-quality nut; and 'Cate,' a commercial cultivar selected in California. Other cultivars that may have potential include recent Hawaiian selections 'Mauka' and 'Pahala,' and California cultivars 'James' and 'Vista.'

RELATIVES The family Proteaceae consists of about 1,000 species. The *Macadamia* genus consists of between 6 and 10 species, but only the 2 species discussed here produce nuts of commercial import.

CLIMATE The tree is intolerant of heavy frost and of prolonged periods of high temperature. Frost damage begins to occur at about 27° F., although a mature tree may withstand brief temperatures as low as 24° or 25° F. Blooms are destroyed at 29° F.

CULTIVATION The macadamia is a low-maintenance tree. It will grow on a wide range of soils, although it requires good drainage. The tree prefers a soil pH of 5.5 to 6.5. Although the macadamia is drought tolerant, fruit production is best where precipitation is evenly distributed throughout the year. In Florida, the tree has few serious pests or diseases. Despite the hard shell, squirrels sometimes develop a fondness for the nut. The tree can be propagated using a whip graft or cleft graft. A branch selected as a scion should be girdled 2 months before it is collected.

HARVEST AND USE The nut matures about 7 months after bloom. It falls to the ground over the cooler months. The nut should be freed from its husk and shell and set out to dry in the shade. After about 2 weeks, the nut can be further dried in an oven set to the lowest temperature setting. It can also be basted in coconut oil and roasted. Whether roasted or raw, the nut makes an excellent snack.

'Cate,' an important commercial cultivar in California, sometimes produces an adequate crop in Florida.

Mamey Sapote

SCIENTIFIC NAME: *Pouteria sapota*
FAMILY: Sapotaceae
OTHER COMMON NAME: Zapote Colorado (Spanish)

Fruiting Calendar

JAN	FEB	MAR	APR	MAY	JUN	JUL	AUG	SEP	OCT	NOV	DEC

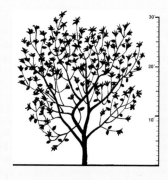

Characteristics

Overall Rating	★★★★
Ease of Care	★★
Taste/Quality	★★★★
Productivity	★★★
Landscape Value	★★
Wind Tolerance	★★★★
Salt Tolerance	★
Drought Tolerance	★★★
Flood Tolerance	★★★
Cold Tolerance	★

The mamey sapote is a rising star among tropical fruit. It has long been popular in Latin America, especially Cuba. Over the last three decades the species has gained favor as a commercial crop in Florida and is occassionally sold in supermarkets. The sweet, soft-textured fruit possesses one of the great flavors of the tropics. It is borne abundantly by a handsome, open tree. The tree is frost sensitive and growth is relegated to coastal areas of south Florida.

Known Hazards

None

138

GEOGRAPHIC DISTRIBUTION The mamey sapote is native to Central America. It is naturalized from southern Mexico through Nicaragua. The origin of the mamey sapote as a commercial crop in Florida can be traced to Eugenio Pantin, a Spanish immigrant and retired electrician. Pantin began to raise Sapoteceae fruit on his Homestead farm. In 1953, he planted Cuban seeds as rootstocks and grafted on them scions taken from a tree growing near the old firehouse on Key West. One of these began to produce fruit of remarkable character. This cultivar, originally called 'Key West,' is now referred to as 'Pantin.' It is the commercial standard in south Florida, accounting for about 75 percent of acreage devoted to the crop. Today, about 500 acres of commercial groves exist in the Redlands area of Miami-Dade County.

Flowers form directly on older wood. This is a branch of the 'Lorito' cultivar.

TREE DESCRIPTION The mamey sapote forms a medium-large tree in Florida, ultimately attaining a height of 40 to 50 feet. In its native habitat it may attain a height of more than 100 feet. In Florida, the tree is briefly deciduous, dropping its leaves in late winter. The crown is irregular, spreading, and open. The foliage is roughly textured and the leaves are tightly clustered near the branch tips. Leaves are simple, obovate, and measure between 5 and 12 inches in length. The underside of young leaves and terminal buds is lightly pubescent.

The attractive foliage of the mamey sapote adds to its appeal as a landscape specimen.

The tree has a short, stout trunk and thick branches. The bark is gray, rough, and somewhat corky. As with many members of the Sapotaceae family, severed twigs exude sticky latex. The rate of growth is moderate and the tree is long lived. It makes a handsome landscape specimen.

The mamey sapote is cauliflorous. Small yellow flowers, measuring about a half inch in diameter, emerge directly from the limbs and form dense clusters. The flowers are bisexual, composed of a 5-celled ovary, 5 stamens, and a 5-lobed corolla.

FRUIT CHARACTERISTICS The typical fruit is elliptic, resembling a football in form. It measures between 4 and 10 inches in length, and may weigh from a few ounces to as much as 4 or 5 pounds. The leathery skin is covered with brown scurf and tiny flecks of corklike grit. The bland exterior of the fruit presents a stark contrast with the brightly colored interior. Flesh color ranges from pink to crimson. The flesh is soft, smooth, fine-grained, and melting. It contains barely noticeable fibers that run longitudinally beneath the skin. The flavor resembles that of a sweet potato or pumpkin, but is sweeter and more refined, containing a hint of almond. Loosely embedded within the center of the flesh are 1 to 4 very large seeds. Most fruit from well-regarded varieties contain a single seed. The seeds are glossy, dark brown, spindle-shaped, and pointed at both ends. Because the fruit require between 14 and 24 months to mature, the current year's crop, the following year's crop, and the flowers may be simultaneously present.

CULTIVARS South Florida has become a secondary center for improvement and distribution of the mamey sapote. Some of the best cultivars for the home garden are listed below.

'Pantin' - This cultivar is the premier commercial variety grown in Florida. Yield is heavy and the tree is upright and well behaved. The fruit are medium size with light red flesh and a single seed. Fruit quality is excellent. Fruit typically ripen in July and August.

'Lorito' - This cultivar, produces a fruit of superb flavor with brilliant red flesh. Fruit are medium in size.

'Lara' - 'Lara' is a red-fleshed selection of exceptional quality. The fruit are medium in size. Fruit ripen from late July through September.

'Pace' - This cultivar, known for its prolific pro-

A mature tree, such as this 'Pantin,' can yield several hundred pounds of delicious fruit.

duction, bears a medium-size fruit of high quality. The flesh is salmon pink. Fruit mature in April and March.

'Viejo' - 'Viejo' produces a small, somewhat rounded fruit with brilliant crimson flesh. Fruit quality is exceptional. Fruit mature in December.

'Abuela' - This cultivar produces a very large fruit with deep red flesh of excellent quality. Fruit mature from October to early December.

'Copan' - This cultivar, a seedling from Cuba, produces fruit of good to excellent quality. The fruit is medium in size, with red flesh. Fruit mature over the late summer.

'Magana' - This cultivar, a seedling from El Salvador, produces an exceptionally large fruit of good quality. The flesh is pink and surrounds a single seed. Fruit development is rapid, with only 12 to 13 months elapsing between flowering and maturity. Fruit ripen in April and May.

Other worthwhile cultivars include 'Cayo Hueso,' 'Chenox,' 'Flores,' 'Francisco Fernandez,' 'Mayapan,' 'Navidad,' 'Piloto,' 'Progreso,' 'Tazumal,' and 'Viejo.'

RELATIVES The mamey sapote belongs to the important Sapotaceae family, which contains about 1,000 species. It is related to the abiu, *Pouteria caimito,* and the canistel, *Pouteria campechiana,* both previously described in this book. It is more distantly related to the sapodilla, *Manilkara zapote,* and the star apple, *Chrysophyllum cainito,* described in subsequent sections. Various Sapotaceae species within the genus *Pouteria* are described under the subsection pertaining to relatives of the canistel.

The green sapote, *Pouteria viridis,* is a close relative of the mamey sapote. This species, native to the medium highlands of Central America, has slightly smaller leaves than those of the mamey sapote. The fruit is ovoid to elliptic, measuring from 2 1/2 to 4 1/2 inches in length. The skin is olive green, turning lighter as the fruit ripens. The flesh is orange-red, closely resembling that of the mamey sapote in appearance, texture, and flavor. The fruit

The cinnamon apple, a relative of the mamey sapote, is productive in south Florida.

140

ripens from January through March—when few other tropical fruits are available. However, without sufficient heat, the fruit often fail to develop full flavor. The green sapote can be successfully grafted onto rootstock of the mamey sapote.

CLIMATE The mamey sapote is tropical in habit. It is planted on the east coast as far north as Jupiter, and on the west coast in the Ft. Myers area. A mature tree will be severely damaged at 28° F and may be killed if the temperature drops to 27° F.

CULTIVATION The mamey sapote requires only moderate maintenance. It will tolerate a fair range of soil conditions, but needs proper drainage. The tree benefits from periodic irrigation, particularly during flowering and periods of drought. Competing weeds should be eliminated from the area beneath the canopy. Frequent light applications of a balanced fertilizer accelerate growth and increase production. Fertilization should be reduced or eliminated over the winter so as not to encourage tender new growth. Chlorosis resulting from iron deficiency is a common problem on the limestone soils of Miami-Dade County. This condition can be corrected through the use of a chelated iron soil drench.

PESTS AND DISEASES Few serious pests or diseases affect the mamey sapote in Florida. Scales, mites, and leafhoppers cause superficial damage to foliage, but rarely require control. The Cuban May beetle attacks the foliage and is a more serious pest. The larva of the sugarcane rootstalk borer feeds on the roots, sometimes inflicting severe damage. The mamey sapote is not a common host of the Caribbean fruit fly.

PROPAGATION The mamey sapote can be grown from seed; however, seedlings are slow to come into production and produce fruit of variable quality. Seeds quickly lose their viability once removed from the fruit. They will germinate in 40 to 100 days, more rapidly if the seed hull is cracked. Seeds should be planted vertically, with the sharply pointed end at the bottom of the hole and the other end level with the surface of the soil.

Grafting is somewhat tedious. It shows the highest rate of success if undertaken in March and April or October and November. Vigorous young shoots are selected as scions. These are girdled a month in advance of the grafting process. Leaves are removed. However, the petioles immediately below the terminal bud are left in place. Veneer grafting and cleft grafting have shown reasonable rates of success. Once grafted, the new plant should be placed within a plastic-bag enclosure to maintain humidity, and should be kept in 50 percent shade. Air layering has also been practiced with some success, especially when undertaken during the late spring or early summer.

HARVEST AND USE Because the mamey sapote does not change color with maturity, care must be taken to harvest the fruit at the proper stage. If the fruit is picked prematurely, it will ripen unevenly or may develop interior mold or other defects. The method for determining maturity is the same as that used for the sapodilla. A small scratch is made through the scurf coating the exterior of the fruit. If the under layer is green, the fruit is immature and may not ripen correctly. If the under layer is tan or red, the fruit is ready to pick. Once it has been harvested, the fruit usually ripens within a week. It is deemed ripe when the skin yields to gentle pressure. Ripe fruit will store in the refrigerator for up to 2 weeks. The mamey sapote can be cut in half and eaten fresh. It is consumed both as a breakfast fruit and as a dessert fruit. It is a prized ingredient for milkshakes, batidos, and ice cream.

The large seed of the mamey sapote, cracked open to reveal the plant embryo within.

Mandarin

SCIENTIFIC NAME: *Citrus reticulata*
FAMILY: Rutaceae
OTHER COMMON NAME: Tangerine, Naranjita or
    Mandarina (Spanish)

Fruiting Calendar

JAN	FEB	MAR	APR	MAY	JUN	JUL	AUG	SEP	OCT	NOV	DEC

Characteristics

Overall Rating	★★★★
Ease of Care	★★★★
Taste/Quality	★★★★
Productivity	★★★
Landscape Value	★★★★
Wind Tolerance	★★★
Salt Tolerance	★★
Drought Tolerance	★★★
Flood Tolerance	★★★
Cold Tolerance	★★★★

The mandarin is a *Citrus* species with many attributes. The fruit is very sweet, with rich, excellent flavor. It is loose skinned and easy to peel. The tree is extremely cold hardy. Several classes and varieties will grow in Florida, and a few of these are capable of surviving and fruiting as far north as the Florida-Alabama border. The mandarin is highly recommended as a dooryard tree throughout Florida.

Known Hazards

Oils in the fruit rind and leaves may occasionally cause dermatitis or photo-sensitivity in sensitive individuals. Occasional spines may be present.

GEOGRAPHIC DISTRIBUTION The mandarin is indigenous to Southeast Asia. The Satsuma, an important class of mandarin, originated in Japan prior to 1600. In the United States the mandarin is grown in Florida, California, Georgia, and the Gulf states. Freezes have caused occasional losses in Alabama, Louisiana, Mississippi, and Texas.

TREE DESCRIPTION The tree is of small to medium size, usually attaining a height between 15 and 20 feet. Leaves are lanceolate, measuring from 3 to 6 inches in length. Petioles are winged. Thorns are present on most varieties. The tree is often grafted on a rootstock of trifoliate orange, which causes some dwarfing. The mandarin is self-fruitful.

FRUIT CHARACTERISTICS The fruit is small to medium in size, typically measuring less than 3 inches in diameter. It is globose, but often has a slight neck and a loose, leathery rind of deep orange coloration. The rind is easily peeled to reveal about a dozen segments of sweet, tender flesh surrounding a hollow core. A few segments may contain seeds, although the fruit is often seedless. The flavor is sweet, rich, aromatic, and easily distinguishable from that of the orange.

CLASSES AND CULTIVARS Mandarins with a reddish-orange peel are often referred to as tangerines. Tangerine cultivars grown in Florida include 'Clementine,' 'Cleopatra,' 'Dancy,' 'Lee,' 'Ponkan,' and 'Sunburst.' Satsuma refers to a class of cold-tolerant mandarins, sometimes treated as a separate species, *Citrus unshiu*. The cultivars include 'Owari' and 'Silverhill.' Other mandarins that do not fall within the tangerine designation or Satsuma class include the cultivars 'Emperor,' 'Kinnow,' and 'Oneco.' 'Ambersweet' and 'Fairchild' are crosses between the 'Clementine' tangerine and the 'Orlando' tangelo. The well-regarded 'Murcott' tangerine may be a cross between a mandarin and an orange.

RELATIVES The family Rutaceae includes all varieties of *Citrus*, and such diverse species as the bael fruit, *Aegle marmelos,* and white sapote, *Casimiroa edulis*. Other *Citrus* species discussed within this book include the grapefruit, *Citrus* x *paradisi;* the lime, *Citrus* spp.; the lemon, *Citrus limon;* and the orange, *Citrus sinensis*.

CLIMATE A mature specimen may withstand a temperature drop to 15° F. However, fruit damage occurs at about 26° F, and defoliation is likely to occur at temperatures of less than 20° F. The tree may be killed by brief exposure to 10° F. Cultivars of the Satsuma class are capable of surviving all but the most severe winters in north Florida. Indeed, in protected locations the Satusuma has fruited as far north as coastal areas of South Carolina. Greatest cold hardiness is attained on specimens grafted on rootstocks of the trifoliate orange.

CULTIVATION The mandarin is a low-maintenance tree and its cultural requirements are similar to those of the orange. It is somewhat drought tolerant, although irrigation enhances fruit set and development. Weeds and grasses should not be permitted to encroach into areas beneath the canopy. Several insect pests attack the mandarin in Florida, including scale insects, various mites, and the citrus leafminer. The mandarin appears to be slightly less susceptible to infection by citrus canker than the sweet orange and grapefruit.

HARVEST AND USE In Florida, fruit ripen in the fall and early winter, from late September through December. The fruit does not hold well on the tree, and will dry out and become puffy unless harvested promptly. Also, once picked, the fruit has poor storage characteristics. The flavor may begin to deteriorate after 10 days at room temperature. Fruit can be used for eating out of hand, juicing, or for any other purpose to which the sweet orange is suited.

The 'Murcott' tangerine is a late variety that is well suited to Florida growing conditions.

Mango

SCIENTIFIC NAME: *Mangifera indica*
FAMILY: Anacardiaceae

Fruiting Calendar

JAN	FEB	MAR	APR	MAY	JUN	JUL	AUG	SEP	OCT	NOV	DEC

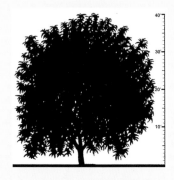

Characteristics

Overall Rating	★★★★★
Ease of Care	★★★
Taste/Quality	★★★★★
Productivity	★★★★
Landscape Value	★★★
Wind Tolerance	★★★
Salt Tolerance	★★
Drought Tolerance	★★★★
Flood Tolerance	★★★
Cold Tolerance	★★

The mango is esteemed for its flavor, productivity, and beauty. It is sometimes referred to as "the apple of the tropics." This description—presumably aimed at showing the popularity of the fruit—does an injustice to the mango's exquisite flavor and exotic appeal. Many consider the mango to be the most delicious fruit that will grow in Florida or any other location. The tree is easy to grow and is reasonably well adapted to conditions in south Florida.

Known Hazards

Exposure to the pollen may cause some to experience respiratory allergies. Exposure to oils in the fruit skin and leaves may cause contact dermatitis. A few individuals are violently allergic to all parts of the plant.

GEOGRAPHIC DISTRIBUTION The mango is native to southern Asia, specifically the region from Myanmar through East India. It has been cultivated for at least 4,000 years, perhaps longer. The mango has been grown in Florida since the 1860s. Early plantings gave rise to the turpentine mango, characterized by small, resinous, fiber-laden fruits. In 1889, the 'Mulgoba' mango was introduced from India. Captain John Haden, of Miami, planted several 'Mulgoba' seedlings. The best of these, first offered for sale in 1912, took his name. 'Haden' produces a fruit of high color, moderate fiber, and fine flavor. While the use of this cultivar has declined, due to its susceptibility to disease, 'Haden' has served as a parent to many important cultivars. During the last century, Florida became a center of diversity and distribution of mango germplasm. A thriving commercial industry existed for several decades. However, increasing land values, foreign competition, high labor costs, and hurricanes conspired to reduce production.

A small flower panicle.

TREE DESCRIPTION The mango is a medium to large evergreen tree. Mature specimens range in height from 25 feet to over 80 feet. The canopy is dense and dome shaped. Leaves are alternate, dark green, lanceolate, measuring between 5 and 12 inches in length. Emerging vegetation is soft textured and may be pink or coppery-red. Growth takes place in successive flushes, preceded by swelling of buds around the branch terminals. A period of active growth is followed by an inactive, hardening-off period. The wood is pliant, but relatively tough. The mango has an extensive root system, which may be anchored by a central taproot. The mango is a tree of moderate to rapid growth. It may live for more than 200 years. The tree makes an attractive landscape specimen, although some cultivars are too large for the average suburban yard. The mango provides dense shade and sheds significant quantities of leaves and other debris. The leaves, wood, sap, and fruit skin contain an irritant that may cause dermatitis and other symptoms in those who are allergic or sensitive.

In Florida, most cultivars flower in February and March. The flowers, which may be yellowish-green, yellow, or orange-red, are densely packed on branched panicles. Hermaphrodite and male flowers coexist on the same panicle. Flowers are tiny, measuring about 1/4 of an inch in diameter. They contain 4 or 5 sepals and a corresponding number of petals. The mango is self-fertile. Various insects serve as pollinating agents.

FRUIT CHARACTERISTICS The fruit is a large drupe, highly variable in form and color. It ranges in size from 2 to 10 inches in length and may weigh from a few ounces to as much as 4 pounds. The skin, which may be green, yellow, orange, pink, red, or purple, is tough, smooth, and inedible. Flesh color ranges from whitish-yellow to deep orange. In some cultivars the taste is mild, sweet, and peach-like. In other selections the flavor is rich, aromatic, and tangy. The flesh encloses a single, flattened, woody stone, bisecting nearly the entire length of the fruit. The stone may be fringed with soft fibers or bristles.

RACES AND CULTIVARS Two races are recognized. The Indian race tends to produce rounded, high-colored fruit. It does poorly in the lowland tropics and is susceptible to anthracnose. The seed is monoembryonic. The Philippines race comes from the Philippines and Southeast Asia. It tends to produce elongated fruit, usually with green or yellow skin. It is tolerant of high humidity. The seed is polyembryonic.

Hundreds of cultivars have been selected and these vary greatly in form and character. The cultivars listed below, while making up only a small faction of those worth planting, produce fruit of exceptional quality and are well suited to the home garden.

'Nam Doc Mai' - 'Nam Doc Mai' is a Southeast Asian mango of excellent quality. The fruit is medium size, dark green, turning yellow when ripe. Fruit are elongated, with a bent apex. The flavor is intensely sweet with delicate, spicy undertones. The texture is fine and melting. Fruit generally ripen in June and July, although sporadic late flowering and fruiting sometimes occurs. The tree is moderately productive. It has good disease resistance but the fruit is prone to splitting under certain conditions.

Compare 'Jehangir,' an Indian mango (left) and 'Sensation,' a Florida selection of primarily Indian heritage (middle), with 'Sia Tong,' a Philippine race mango (right).

'Glenn' - This cultivar is frequently recommended as a mango for home planting. It originated in Miami in the 1940s. The fruit is medium size, yellow with an orange blush. It is of excellent eating quality, but has poor storage characteristics. It ripens in June and July. The tree fruits abundantly and consistently with minimal care.

'Carrie' - This mid-season variety originated in Boynton Beach around 1937. The small to medium, green-skinned fruit is slightly oblong in form. Although the fruit is not attractive from the outside, it excels in eating quality. It is sweet, aromatic, and fiberless. 'Carrie' fruits in June and July. The tree is compact, perfect for the homeowner with a small yard.

'Mallika' - 'Mallika' is a cross between two Indian dessert mangoes. It was introduced to Florida from India in 1978. The fruit is small to medium, yellow, and slightly elongated. 'Mallika' produces fruit of sweet, rich flavor, which improves with several days of storage at room temperature. Fruit ripen in June and July. The tree is sometimes reported as dwarf, but has proven to be vigorous and fairly large in Florida.

'Neelum' - This Indian dessert mango, a parent of 'Mallika,' deserves inclusion in this list both as a result of its excellent eating quality, and because of its late maturation. The fruit ripen from September into early October. The tree is relatively compact.

'Graham' - This cultivar was introduced from Trinidad in 1932. The fruit is medium-large and rich and spicy in flavor. Ripening takes place from July through August.

'Rosigold' - This cultivar was recently rediscovered in a Miami garden. The fruit is slightly elongated, oval, and small. The key attribute of 'Rosigold,' apart from its outstanding flavor, is that it fruits exceptionally early, in April and early May. The fruit may be susceptible to internal breakdown under certain conditions.

'Cogshall' - This variety originated on Pine Island, Florida, in the 1940s. The fruit is medium size, yellow-orange, sometimes with a red blush. It is fiberless, with a rich, spicy flavor. The tree is relatively small. A midseason mango, 'Cogshall' ripens in June and July.

'Edward' - 'Edward' originated in the Miami area in the 1920s. Production is sparse but reasonably consistent. Many regard 'Edward' as the world's finest-flavored mango. Fruit ripen from late May to early July. The fruit is medium size, yellow-green with a slight pink blush.

Other cultivars well suited to home planting include 'Carabao,' 'Cushman,' 'Dot,' 'Florigon,' 'Ice Cream,' 'Julie,' 'Kent,' 'Keitt,' 'Okrong,' 'Sri Chinook,' and 'Zill.'

RELATIVES The family Anacardiaceae contains several important fruiting species. Described within this book are the ambarella, *Spondias dulcis;* cashew, *Anacardium occidentale;* and mombin, *Spondias* spp. The family also includes the dragon plum, *Dracontomelum* spp.; imbú, *Spondias tuberosa;* kaffir plum, *Harpepbyllum caffrum;* marula, *Sclerocarya birrea;* and pistachio, *Pistacia vera.* Anacardiaceae species native to Florida include such dermatitis-inducing plants as poisonwood, *Metopium toxiferum;* poison sumac, *Toxicodendron vernix;* and poison ivy, *Toxicodendron radicans.*

More closely affiliated with the mango is the gandaria or ma prang, *Bouea macrophylla,* native to Malaysia, Sumatra, and Java. This widely cultivated species resembles the mango in its habit of growth. It bears small mangolike fruit. 'Ramania Pipit,' 'Ramania Tembaga,' and 'Wan' are respected cultivars. In Florida, the gandaria bears in April and May.

The genus *Mangifera* includes about 70 species, mostly native to Southeast Asia. At least 25 produce edible fruit, including the binjai mango, *Mangifera caesia,* which bears fruit with white flesh and agreeable flavor; the horse mango, *Mangifera foetida,* which bears large fruit of inferior quality; the kuwini mango, *Mangifera odorata,* which bears green-skinned fruit with sweet, aromatic flesh; and the mempelam bemban, *Mangifera pentandra,* which bears small fruit with greenish-yellow skin and sweet orange flesh.

CLIMATE The mango requires a tropical or subtropical climate. It depends on a pronounced dry season to stimulate flowering. The tree can be severely damaged when the temperature falls below 28° F. However, a grove of mature trees in Stuart survived the freeze of 1989—after exposure to a temperature of 26° F on two successive nights. Mature trees in Ft. Pierce, 20 miles to the north, were killed. The mango will grow as far north as Merritt Island, on the east coast, and Sarasota, on the Gulf coast.

CULTIVATION The mango is a low-maintenance tree and requires little in the way of care once it is established. The tree prefers a soil pH of 5.5 to 7.5. It will grow in nearly any well-drained soil. The tree is drought tolerant. A young tree should be watered twice a week for the first 2 months after planting, and once a week for the next 2 months. Thereafter, the tree rarely requires supplemental irrigation. A young tree should receive frequent light feedings of a 12–3–12 fertilizer from February through October. A mature tree will benefit from 2 applications of a balanced fertilizer, the first in late winter upon the emergence of flower panicles, the second immediately after harvest.

PESTS AND DISEASES Anthracnose is by far the most serious disease affecting the mango in Florida. This fungal disease attacks various parts of the plant at nearly every stage of the fruiting cycle. It is most prevalent during periods of wet weather. To prevent crop failure in susceptible cultivars, fungicide must be applied on a biweekly basis, from just before the emergence of blooms until harvest. Other diseases include powdery mildew, stem end rot, and internal breakdown. Few pests bother the mango in Florida. Scale insects are usually present to some extent, but rarely require control. Damaged or overripe fruit are occasionally invaded by the Caribbean fruit fly.

PROPAGATION The mango will grow readily from seed. However, quality is variable, particularly among seedlings of the monoembrionic Indian race. Grafting is the primary method of vegetative reproduction. Veneer grafting and cleft grafting are both carried out with near-equal success. The best scions are shoots that are well hardened off with swollen terminal buds. These may be defoliated 2 weeks prior to use.

HARVEST AND USE A 15-year-old tree of a good cultivar may be expected to produce 250 or more fruit annually. Most fruit ripen over the summer and fall, from 120 to 180 days after flowering. The stem of a ripe fruit will pull from the plant with minimum pressure. A pole with a V-shaped blade may be used to harvest fruit near the top of the tree. The mango can be used in chutneys and relishes, and for flavoring various dishes. However, the fruit may be at its best when it is sliced and eaten out of hand.

'Nam Doc Mai,' perhaps the premium Philippine race mango, offers a sensational taste enclosed in an unassuming exterior.

Mayhaw

SCIENTIFIC NAME: *Crataegus* spp.
FAMILY: Rosaceae
OTHER COMMON NAME: May Hawthorn

Fruiting Calendar

JAN	FEB	MAR	APR	MAY	JUN	JUL	AUG	SEP	OCT	NOV	DEC

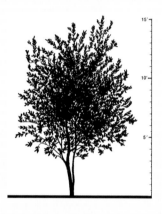

Characteristics

Overall Rating	★★★
Ease of Care	★★★★
Taste/Quality	★★
Productivity	★★★★
Landscape Value	★★★★
Wind Tolerance	★★★★
Salt Tolerance	★
Drought Tolerance	★★★
Flood Tolerance	★★★★★
Cold Tolerance	★★★★★

The mayhaw, native to north Florida, produces a small pome that has been used in traditional Southern recipes since antebellum days. The tree is attractive, productive, and will grow on ground too moist for other fruit crops. The fruit is the main ingredient in mayhaw jelly—often touted as the world's best jelly. For those residing in north and north central Florida the mayhaw makes a worthwhile addition to the home landscape.

Known Hazards

Sharp spines are present on most varieties.

GEOGRAPHIC DISTRIBUTION The mayhaw inhabits lowland areas of Florida, Mississippi, Louisiana, Texas, and Georgia. It is an important part of Southern folk heritage. Will McLean (1919–1990), a folk singer and member of the Florida Artists Hall of Fame, spoke of wading into a slough "to gather a brimming hat full of wild mayhaws from a tree standing in the water. They knew the mayhaw at its pink and white blossom time, the jelly of its fruit, and as now, the luxury of tasting its tarty sweetness." The mayhaw is gradually moving from the swamp into the commercial orchard. The future of the species appears bright.

TREE DESCRIPTION The mayhaw is a small deciduous tree. It may eventually attain a height of about 25 feet. Leaves are about 2 inches in length. They are deeply toothed, ovate to elliptical, broad toward the base and narrow toward the apex. Most cultivars have spines on smaller branches. The tree is relatively long-lived, sometimes bearing fruit for more than 50 years. It blooms from late February through March. Flowers have 5 white petals and 15 to 20 pink stamens.

Leaves of a wild north Florida mayhaw (top) are similar, but not identical, to those of the commercial cultivar 'T.O. Superberry' (bottom).

FRUIT CHARACTERISTICS The fruit is a small pome, measuring from 1/2 to 1 inch in diameter. In outward appearance it resembles a cross between a crabapple and a cranberry. The skin is glossy. The flesh is juicy, highly aromatic, tart, and tangy.

SPECIES AND CULTIVARS The mayhaw is made up of 3 species: *Crataegus opaca*, the western species; *Crataegus rufula*, the central species; and *Crataegus aestivalis*, the eastern species. At least 50 cultivars have been selected. Superior choices for planting in Florida include 'Texas Superberry,' a heavy producer with low chilling requirements that bears large, bright red fruit with pink flesh; 'Big Red,' a late-blooming cultivar that bears large fruit of good quality; 'Saline,' a late-blooming cultivar that bears large, firm fruit with red skin and orange-pink flesh; and 'Texas Star,' a late-blooming cultivar that bears large orange-red fruit with yellow flesh. Other reputable cultivars include 'Spectacular,' 'T.O. Superberry,' 'Red and Yellow,' 'Superspur,' 'Perfection,' 'Royal Star,' 'Maxine,' 'Turnage 57,' and 'Turnage 88.'

RELATIVES The mayhaw is a member of the Rosaceae or Rose family. Other pome species include the apple, *Malus domestica;* loquat, *Eriobotrya japonica;* and pear, *Pyrus* spp., reviewed in separate sections. The mayhaw belongs to the *Crataegus* or hawthorn genus. At least 10 species within the genus are native to Florida. None of these, apart from the mayhaw, bear fruit of particular merit. More distant native relatives include the common serviceberry, *Amelanchier arborea,* and the southern crabapple, *Malus anggustifolia.*

CLIMATE The mayhaw can survive temperatures of -20° F, although late frosts sometimes interrupt flowering. Some varieties have a low chilling requirement and will fruit in Orlando. The southern limits of production have not been adequately established.

CULTIVATION The mayhaw has moderate maintenance requirements. Periodic irrigation is needed when the tree is grown on well-drained soil. It can tolerate flooding and moist soils. Insect pests include the plum curculio, the apple borer, white flies, the hawthorn lace bug, and scale insects. The tree is susceptible to quince rust, a fungal disease. While the mayhaw can be grown from seed, better cultivars are grafted using a cleft graft or whip and tongue graft.

HARVEST AND USE The fruit typically ripens from late April through May. A 20-year-old tree may yield between 15 and 30 gallons. To harvest the fruit, a tarpaulin is spread beneath the canopy and the tree is given a vigorous shake. The mayhaw is used in jellies, marmalades, sauces, preserves, condiments, syrups, and juices.

Mombin

SCIENTIFIC NAME: *Spondias* spp.
FAMILY: Anacardiaceae
OTHER COMMON NAME: Hog Plum, Spanish Plum,
    Ciruela or Jacote (Spanish)

Fruiting Calendar

JAN	FEB	MAR	APR	MAY	JUN	JUL	AUG	SEP	OCT	NOV	DEC

Characteristics

Overall Rating	★★
Ease of Care	★★★
Taste/Quality	★★
Productivity	★★★★
Landscape Value	★★★
Wind Tolerance	★★★★
Salt Tolerance	★★★
Drought Tolerance	★★★★
Flood Tolerance	★★★
Cold Tolerance	★

The mombin, a distant relative of the mango, bears a small, aromatic, plumlike fruit. Quality ranges from fair to good. The species is very productive and is commonly planted in many tropical regions. However, the tree is intolerant to winter cold in Florida and is limited to southern coastal areas. The ripe fruit is susceptible to infestation by the Caribbean fruit fly.

Known Hazards

Contact with leaves or sap may cause dermatitis in some individuals.

150

GEOGRAPHIC DISTRIBUTION The mombin is native to the American tropics, from southern Mexico to northern South America. While it has been commercialized to some degree, it is most often planted as a dooryard crop.

TREE DESCRIPTION The mombin is a medium to large deciduous tree. Leaves are alternate and pinnately compound, with each leaf containing between 5 and 21 leaflets. The leaflets measure 1 to 1 1/2 inches in length.

FRUIT CHARACTERISTICS The fruit, although variable in shape, resembles a small plum. It ranges from 1 to 2 1/4 inches in length. The skin is shiny and thin, with moderate toughness and resilience. Depending on the species and strain, the skin color may be yellow, orange, red, or purple. The flesh is yellow, firm, and slightly fibrous in texture. The flavor is aromatic, sub-acid, somewhat musky, and plumlike. Embedded in the center of the flesh is a large, irregular stone, which contains several seeds or seed remnants.

SPECIES Two species exist: the red mombin, *Spondias purpurea,* and the yellow mombin, *Spondias mombin.* The red mombin forms a spreading tree. Small, reddish-purple flowers are born on compact panicles before the tree leafs out in the spring. The fruit of the red mombin is considered superior to that of the yellow mombin. The yellow mombin is more upright in form. Small, white flowers appear after the emergence of foliage in the spring. Fruit are born in dangling clusters. While many cloned varieties exist, few are available in Florida and most trees are propagated on an informal basis.

RELATIVES The mombin, as a member of the Anacardiaceae family, shares a distant affinity with the mango, *Mangifera indica;* pistachio, *Pistacia vera;* and cashew, *Anacardium occidentale.* The *Spondias* genus contains a number of fruiting species. The ambarella, *Spondias dulcis,* is widely grown and is discussed within its own section. The imbú, *Spondias tuberosa,* is a small tree that produces an oviod fruit of good quality. Native to northeastern Brazil, it prefers near-arid conditions. Unfortunately, attempts to grow the imbu to Florida have not been successful. Other less-familiar fruiting species include *Spondias pinnata,* of tropical Asia, and *Spondias borbonica,* of Mauritius and Reunion Island.

CLIMATE The mombin is tropical in its requirements and is severely damaged when temperatures fall to 28° F. In Miami-Dade County's Fruit and Spice Park, the mombin grows well and fruits heavily.

CULTIVATION The mombin is a durable tree, capable of growing in poor quality soil. It is drought tolerant and requires little care. The fruit, especially when overripe or damaged, may be infested with larvae of the Caribbean fruit fly. The red mombin does not bear viable seed. The yellow mombin can be grown from seed, but there is little advantage in doing so. Both species are almost universally reproduced from large cuttings. Branches are severed and placed upright in containers or in the ground. These quickly root and begin to grow. Within its native range, the mombin is often used to form living fence posts. These are connected with barbed wire to control the movement of cattle.

HARVEST AND USE In Florida, the mombin typically fruits during the summer. The fruit may be eaten raw, made into jelly, or used in tarts or pies. The fruit can be preserved for future consumption in various ways. It can be boiled for 5 minutes in a brine solution and dried. It can also be pickled in vinegar and candied in syrup.

The tiny flowers, emerging fruit, and developing fruit of the mombin.

Mulberry

SCIENTIFIC NAME: *Morus* spp.
FAMILY: Moraceae

Fruiting Calendar

JAN	FEB	MAR	APR	MAY	JUN	JUL	AUG	SEP	OCT	NOV	DEC

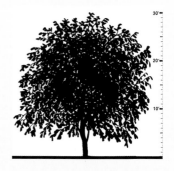

Characteristics

Overall Rating	★★★★
Ease of Care	★★★★
Taste/Quality	★★★★
Productivity	★★★
Landscape Value	★★★
Wind Tolerance	★★★★
Salt Tolerance	★★
Drought Tolerance	★★★★
Flood Tolerance	★★★
Cold Tolerance	★★★★★

The mulberry is an underrated fruit with great potential as a dooryard crop. The tree produces prodigious quantities of delicious, berrylike fruit. It is cold hardy, drought tolerant, fast growing, and requires little care. It establishes quickly and bears at a young age. However, the fruiting season is short, the fruit leaves an indelible stain, and the Caribbean fruit fly is an occasional pest in southern parts of the peninsula.

Known Hazards

The white mulberry is an invasive exotic.

152

GEOGRAPHIC DISTRIBUTION The white mulberry originated in China. It is considered an invasive exotic in the United States. The black mulberry is native to western Asia and the Middle East. The red mulberry is native to the eastern half of the United States, including Florida.

TREE DESCRIPTION Notwithstanding the popular nursery song, the mulberry is not a bush. It is a medium deciduous tree. Leaves are alternate, broad and ovate, but variable in shape. Leaf margins are roughly serrated and can be notched or lobed. The trunk may exceed 2 feet in diameter and is covered with thin gray or grayish brown bark. Flowers are pendulous, light green catkins.

FRUIT CHARACTERISTICS The collective fruit, which may be red, green, white, purple, or black in color, is composed of the swollen ovaries of numerous tiny flowers. In better cultivars, the flavor is sweet, but with a hint of tartness and acidity. The fruit stem penetrates the length of the fruit.

SPECIES AND CULTIVARS Three species are commonly grown. The native red mulberry, *Morus rubra,* bears reddish or black fruit of very good quality. The black mulberry, *Morus nigra,* bears fruit of exceptional quality. Leaves tend to be smaller than those of other species. The fruit is almost always black. The white mulberry, *Morus alba,* has white buds that emerge in early spring, usually before those of the red mulberry and well before those of the black mulberry. The fruit, which may be black, purple, or white, tends to be sweet but insipid. A number of excellent mulberry cultivars are available. Those varieties well suited to Florida, include 'Tice,' 'Pakistan,' 'Tehama,' 'King White Pakistan,' 'Shangri-La,' 'Bachuus Noir,' 'Red Gelato,' 'Black Persian,' 'Illinois Everbearing,' and 'White Dove.'

RELATIVES The mulberry is a member of the Moraceae family, which includes the che, *Cudrania tricuspidata;* the jackfruit, *Artocarpus heterophyllus;* and the fig, *Ficus carica,* profiled within separate sections of this book.

CLIMATE The mulberry is primarily a temperate fruit tree and is not damaged by the lowest temperatures that occur in Florida.

CULTIVATION The tree requires very little maintenance. It is tolerant of poor soil and appears to thrive in the infertile sandy soil found across much of the Florida peninsula. The tree is drought tolerant and, once established, rarely requires irrigation. It is moderately resistant to high winds. In most locations, the tree grows and fruits adequately without fertilization. To avoid stains from fallen fruit, the mulberry should not be planted over driveways, patios, or walkways. This problem can be avoided by choosing a light-fruited cultivar such as 'Tehama' or 'King White Pakistan.' Light pruning may be undertaken when the tree is young, to establish a strong framework of main branches. Once the tree matures it should not be pruned except to remove dead wood or crossing limbs. Wounds caused by the removal of major branches are slow to heal. While few serious pests or diseases affect the tree, the Caribbean fruit fly sometimes infests the fruit in south Florida. The tree is easily cloned through softwood cuttings and by T-budding sprigs.

HARVEST AND USE Most cultivars ripen over the spring and early summer. The mulberry is excellent eaten out of hand and can be substituted for the blackberry in recipes.

The 'King White Pakistan' mulberry turns light green, then white, as it ripens.

Muscadine Grape

SCIENTIFIC NAME: *Vitis rotundifolia*
FAMILY: Vitaceae
OTHER COMMON NAMES: Scuppernong (bronze-skinned
    varieties), Bullet (dark-skinned varieties), Uva
    Muscadine (Spanish)

Fruiting Calendar

JAN	FEB	MAR	APR	MAY	JUN	JUL	AUG	SEP	OCT	NOV	DEC

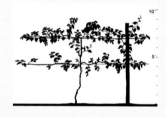

Characteristics

Overall Rating	★★★★★
Ease of Care	★
Taste/Quality	★★★★★
Productivity	★★★★
Landscape Value	★★
Wind Tolerance	★★
Salt Tolerance	★
Drought Tolerance	★★★
Flood Tolerance	★★★
Cold Tolerance	★★★★★

The muscadine grape is a native treasure. The fruit is excellent—deserving wider cultivation and consumption. It is several times the size of the average commercial grape, but is borne in small groups rather than tight bunches. The flavor is aromatic, spicy, sweet, and musky. The plant is relatively hardy and resists several diseases that afflict bunch grapes. However, it requires trellising, pruning, and other tasks associated with viniculture. The reward is well worth the effort.

Known Hazards

None

154

GEOGRAPHIC DISTRIBUTION The muscadine grape is native to the southeastern United States. It ranges from Delaware, south to Florida, and west to Texas. The species can be found growing wild in nearly every county of Florida. Native Americans enjoyed this grape long before the arrival of the Europeans. In Florida, it was consumed and made into wine by early Spanish settlers. The name muscadine is thought to have developed when European settlers—familiar with the muscat grape of France—used a derivation to refer to the sweet, musky grapes they found growing in the American South.

The University of Florida, University of Georgia, University of North Carolina, and Florida A&M University have all been involved in research or breeding programs. Several small commercial operations exist in north Florida. There has been a decline in acreage over the past two decades, perhaps indicating that the muscadine is not yet ready to transcend its status as a regional crop. Florida has at least 10 small wineries, scattered throughout the state. Many of these rely in part on the muscadine grape.

PLANT DESCRIPTION The muscadine grape grows on a large woody vine. The vine is vigorous and climbing. It clings to supporting structures with powerful, coiled tendrils. Leaves are simple, alternate, and measure from 3 to 5 inches in diameter. They are rounded, with coarsely notched or unevenly serrated edges. The tiny green flowers, which form on tight panicles, may be male, female, or perfect. For the most part, the muscadine is dioecious, bearing male and female flowers on different plants. Female flowers require cross-pollination from a male or perfect flower in order to set fruit. Therefore, female plants are usually inter-planted with self-fertile varieties. Insects play an important role in pollination.

The buds have not yet opened on this flower panicle.

FRUIT CHARACTERISTICS The fruit is a thick-skinned berry, usually borne in small, loose clusters on new growth. It is much larger than the average bunch grape, measuring up to 1 1/2 inches in diameter. It ranges in color from black to purple to red to bronze. The average fruit contains 3 or 4 seeds. The flavor is musky and is sometimes very sweet. The muscadine has a distinct spiciness that sets it apart from typical bunch grapes. Those who are accustomed to muscadine grapes often consider them superior to the "bland" grapes found in supermarkets. Some varieties leave a slight tingling or burning sensation around the edges of the mouth, which in no way detracts from the eating quality. The fruit typically ripens from July through October.

CULTIVARS A surprising number of cultivars have been selected over the years. Cultivars recommended for dooryard production in Florida are listed below.

'Supreme' - This early- to mid-season cultivar bears very large black fruit, sometimes measuring more than 1 1/2 inches in diameter. The fruit is sweet and of excellent flavor. 'Supreme' has fair cold tolerance. Although this cultivar is not self-pollinating, production is very high.

'Darleen' - 'Darleen' is a very large-fruited variety. The fruit is bronze in color, sometimes with a red blush. The quality is excellent with sweet, melting flesh. Yield is high, with the fruit maturing mid-season. It is not self-pollinating.

'Summit' - This cultivar has the advantage of being self-fruitful. It bears medium-size, bronze-skinned fruit. The fruit is sweet and of very high quality. The skin is thin. 'Summit' has medium cold tolerance and will produce in south Florida. 'Summit' is an early- to mid-season variety.

'Black Beauty' - This cultivar bears large, very sweet fruit of excellent quality. It is not self fruitful and requires a pollinator. Yield is moderate. Fruit ripens mid-season. 'Black Beauty' has fair cold hardiness.

'Nesbitt' - This cultivar is self-fruitful and ripens from September through October. It has low chilling requirements and is suitable for growing in south Florida. The fruit is black, of medium sweetness, and of good quality. Size is medium, with the fruit typically measuring just over an inch in diameter. Production is high.

'Supergate' - 'Supergate' is an early-season cultivar that bears medium-size black fruit. The sugar content of the fruit is very high. 'Supergate' requires a pollinator. Yield is moderate.

'Doreen' - This bronze-fruited variety bears very sweet, medium-size fruit of excellent quality. It is self-fruitful. 'Doreen' is a late season cultivar, with good cold hardiness. It is recommended for dooryard planting in north and central Florida.

Muscadine Grape

'Fry' - This cultivar is somewhat susceptible to cold injury in extreme north Florida. However, the fruit are very large—nearly 1 1/2 inches in diameter—and are of excellent quality. 'Fry' has low chilling requirements and will grow and fruit in south Florida. Skin color is bronze.

'Sweet Jenny' - 'Sweet Jenny' is an early- to mid-season variety that produces very large, very sweet fruit. Skin color is bronze. Yield is moderately high. This cultivar is not self-pollinating.

'Alachua' - 'Alachua' bears medium-size black fruit of good quality and medium sweetness. It is self fertile and very productive. The skin is relatively thin, making it suitable for fresh consumption. 'Alachua' is a mid-season cultivar with fair cold hardiness.

Other well regarded cultivars include 'Albemarle,' 'Carlos,' 'Cowart,' 'Dixie Red,' 'Dixieland,' 'Higgins,' 'Janet,' 'Jumbo,' 'Magnolia,' 'Noble,' 'Pam,' 'Pineapple,' 'Scarlet' 'Sterling,' 'Sugargate,' 'Tara,' and 'Welder.' A variety recently developed by the University of Florida, designated 'AAII-68,' produces a good quality fruit in easily harvested bunches.

RELATIVES The Vitaceae or Grape family is composed of climbing vines. It is estimated to contain about 700 species in 10 to 12 genera. Other grape species native to Florida include the summer grape, *Vitis aestivalis,* widely distributed across the state; the graybark grape, *Vitis cinerea,* of the western panhandle; the Florida grape, *Vitis cinerea* var. *Floridana,* widespread through peninsular Florida; the catbird grape, *Vitis palmata,* of north Florida; the Caloosa grape, *Vitis shuttleworthii,* of south and central Florida; and the frost grape, *Vitis vulpina,* of north Florida and west central Florida. These native species are tough but tend to produce small, acidic fruit.

The European grape, *Vitis vinifera,* employed for centuries as a wine and table grape, is only marginally adapted to Florida's subtropical climate. Moreover, its susceptibility to Pierce's disease, which is omnipresent in the environment, makes cultivation all but impossible. Hybridization of European grapes with native grapes has produced bunch grape cultivars suitable for growth in Florida. These include 'Blanc Du Bois,' 'Blue Lake' 'Daytona,' 'Lake Emerald,' 'Orlando Seedless,' 'Stover,' and 'Suwannee.' 'Southern Home' is a self-fruitful muscadine x bunch grape hybrid developed in Florida.

CLIMATE The muscadine grape requires a warm-temperate to subtropical climate. It thrives in summer heat and humidity and is well adapted to growing conditions in Florida. It can tolerate tempera-

ture drops to 10° F or lower and is therefore not likely to be injured by winter cold in Florida.

CULTIVATION The muscadine, like most grape vines, is a relatively high-maintenance plant. It will adapt to various soil types and makes adequate growth in the sandy soils prevalent across much of the Florida peninsula. Proper drainage is important, and heavy clay and hardpan should be avoided. Optimal soil pH is 6.0 to 6.5.

The vine should be started in the early spring, once the danger of frost has passed. Unless the

If properly trellised, the vine will support heavy loads of fruit.

plant is root bound, care should be taken not to disturb the root ball at planting. The muscadine should receive full sun and should be spaced no closer than 10 feet apart. Young plants should be periodically irrigated and should receive water on a weekly or biweekly basis. Established plants are fairly drought tolerant.

Plants establish more readily if peat is added to the soil at the time of planting. The muscadine responds well to fertilization. Young plants should receive several light applications of a balanced fertilizer over the course of the warmer months. Mature vines should receive multiple applications totaling about 5 pounds annually.

A 2- or 3-wire trellis system is critical to production. Wires should be run between vertical posts. A stake should be inserted next to each young plant, and the main leader should be fastened to this stake to encourage vertical growth. All lateral shoots should be removed until the lower trellis wire is reached, wherein two shoots should be selected for lateral growth. Two additional shoots are permitted to branch out into lateral arms or "cordons" along the central wire. When the vertical leader reaches the top wire, it should be pinched to encourage two additional cordons. Once the cordons have stretched the length of the wire, they are permitted to develop side shoots or spurs. Fruit production takes place on shoots that originate from these spurs.

Pruning is required to stimulate the production of new shoots, to keep the vines in bounds and to ensure productivity. Once formed, the trunk and main laterals remain in place. The spurs must be cut back annually leaving only 2 or 3 buds for fruit production during the next season. Shoots that spring from older wood rarely set fruit and should be eliminated. Pruning is undertaken during the winter, while the vine is dormant

PESTS AND DISEASES The muscadine grape is resistant to Pierce's disease and phylloxera. Bitter rot, a fungal disease, sometimes affects fruit in Florida. As the fruit approaches maturity it turns black and the skin is covered with spore-bearing structures. Black rot, another fungal infection, causes black spots on the fruit and brown spots on the leaves. Ripe rot causes brown lesions punctuated by orange, spore-bearing structures. Birds are notorious pests and may account for serious losses.

PROPAGATION The muscadine can be reproduced by ground layering and by softwood cuttings rooted under intermittent mist. Selected cultivars can be grafted onto seedling muscadine rootstocks.

HARVEST AND USE Annual yield from a single vine can exceed 25 pounds of fruit. Most varieties ripen in August and September, although a few ripen as early as July or as late as October. Because bunches do not mature at once, the grapes must be picked individually. Ripe fruit tends to abscise from the stem and fall to the ground. The fruit stores well and may be refrigerated for short periods. It is excellent eaten out of hand. It can also be juiced or used to make jelly and preserves.

The fruit of the muscadine grape may not have the blemish-free exterior of store-bought grapes, but the flavor is rich, sweet, and spicy.

Ogeechee Lime

SCIENTIFIC NAME: *Nyssa ogeechee*
FAMILY: Nyssaceae
OTHER COMMON NAME: Ogeechee Tupelo

Fruiting Calendar

JAN	FEB	MAR	APR	MAY	JUN	JUL	AUG	SEP	OCT	NOV	DEC

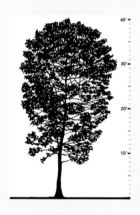

Characteristics

Overall Rating	★★★
Ease of Care	★★★★
Taste/Quality	★★
Productivity	★★★★
Landscape Value	★★★★
Wind Tolerance	★★★
Salt Tolerance	★
Drought Tolerance	★★
Flood Tolerance	★★★★★
Cold Tolerance	★★★★★

The Ogeechee lime is an obscure but intriguing native species. It is a fruit tree with untapped potential. The tree is beautiful and is one of only a handful of fruiting species that will grow in wet locations. The fruit, while not recommended for eating out of hand, has several culinary uses. The Ogeechee lime is highly recommended as a dooryard tree for those living in north and north central Florida who are willing to try something different.

Known Hazards

None

158

GEOGRAPHICAL DISTRIBUTION The Ogeechee lime is native to north Florida, south Georgia, and coastal regions of Georgia. It is often closely associated with rivers and moist areas, where it sometimes forms thicketlike stands. In north Florida, it is prevalent along the banks of the Apalachicola, Ochlockonee, and Suwannee rivers. It occurs naturally at least as far south as Alachua County.

TREE DESCRIPTION The Ogeechee lime is a medium, deciduous tree, capable of attaining a height of about 40 feet. The crown is somewhat pyramidal when young, of medium density and of fine to medium texture. The alternate, dark green leaves are elliptic to obovate, measuring 4 to 6 inches in length. The leaves are usually entire, but may have a few irregularly spaced teeth along one or both margins. In the fall, the leaves take on an array of brilliant colors. Inconspicuous clusters of small white or greenish-white flowers emerge in the spring, primarily in April. The species is generally self-pollinating. Some trees produce both perfect and male flowers. However, others bear only male flowers and will not set fruit. Nectar from the Ogeechee lime and related species is the source of "tupelo honey," regarded by many as the world's finest honey. Growth of the tree is moderately fast. It may begin to bear fruit in as little as 3 years.

FRUIT CHARACTERISTICS The fruit is an oblong, berrylike drupe, about 1 1/2 inches in length. It is green when immature, blushing scarlet upon ripening. The flesh is juicy and contains a single stone surrounded by about 10 small papery wings. The flavor is pleasant, although the fruit is usually too sour to be eaten out of hand.

RELATIVES The Nyssaceae family is tiny, made up of only 2 genera, which contain a total of about 10 species. It is closely associated with the Cornaceae or Dogwood family and some authorities place it within that larger family. The Nyssa or tupelo genus contains about 5 species native to the southeastern United States. The water tupelo, Nyssa aquatica, is often confused with the Ogeechee lime and produces a similar, but smaller, fruit. The species occurs from Virginia to Texas. The tupelo or black gum, Nyssa sylvatica, has even wider distribution and is present throughout most of the eastern United States.

CLIMATE The Ogeechee lime is not affected by winter cold in Florida. It can probably be grown in areas well south of its natural range. One report has the tree occurring in Hillsborough County.

CULTIVATION The Ogeechee lime is a low-maintenance tree. It will grow in many types of soil and will tolerate damp soil and significant flooding. When grown on well-drained soil, it must be watered until well established. Thereafter, it requires irrigation only during periods of drought. It can be grown in full sun or light shade, but will not tolerate deep shade. The species does not appear to be affected by any serious pests or diseases. Propagation is entirely by seed and little if any effort has been made to select superior strains.

HARVEST AND USE The Ogeechee lime ripens during the fall, usually in September just prior to the time of leaf drop. Fruit production is prolific and a single tree is capable of bearing several hundred pounds of fruit. The fruit can be shaken from individual branches onto tarpaulins. It eventually falls to the ground for easy harvest. The Ogeechee lime is rarely eaten out of hand. It is primarily used as a substitute for lime juice, as a flavoring for drinks, and as an ingredient in sauces, chutneys, and marinades. It can also be made into preserves.

Olive

SCIENTIFIC NAME: *Olea europaea*
FAMILY: Oleaceae
OTHER COMMON NAME: Olivo (Spanish)

Fruiting Calendar

JAN	FEB	MAR	APR	MAY	JUN	JUL	AUG	SEP	OCT	NOV	DEC

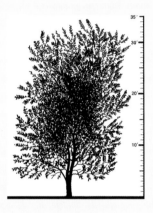

Characteristics

Overall Rating	★★
Ease of Care	★
Taste/Quality	★★★
Productivity	★★
Landscape Value	★★★★★
Wind Tolerance	★★★★★
Salt Tolerance	★★★
Drought Tolerance	★★★★★
Flood Tolerance	★★★
Cold Tolerance	★★★★

The olive, a fruit enjoyed from antiquity, is sometimes planted as an ornamental in the southeastern United States. The tree is hardy and beautiful. Unfortunately, in Florida, fruit production is often sparse as a result of high humidity and lack of winter chill. While the olive may never assume the import in Florida that it enjoys in California, it should be regarded as a valuable if occasional dooryard crop.

Known Hazards

The fruit is inedible until it has been processed.

GEOGRAPHIC DISTRIBUTION The olive is native to the eastern Mediterranean region. It has been domesticated for at least 3,000 years. In the United States, commercial production is centered in California. In 1831, the Marquis de Lafayette attempted to establish olive groves near Tallahassee. The project did not succeed.

TREE DESCRIPTION The olive is a small evergreen tree. It grows to a height of about 30 feet. The crown is symmetrical and finely textured. Leaves are opposite, leathery in texture, and gray-green in color. They are oval, with entire margins, measuring between 1 1/2 and 3 inches in length. The tree has a slow rate of growth and may live for more than 500 years. The olive should not be planted near drives, walks, or patios, as fallen fruit will stain porous surfaces. The tree blooms in the spring, producing thousands of clusters of small, fragrant, cream-colored flowers. These may be either perfect or staminate. The tree is almost exclusively wind pollinated. Most cultivars are somewhat self fertile, although fruit set increases with cross-pollination between cultivars.

FRUIT CHARACTERISTICS The fruit is a small drupe with smooth, thin skin. The skin is green in most varieties. The shape is oval, measuring from 1/2 to 1 1/2 inches in length. Each fruit contains a single, brown, elongated seed. The flesh of the fresh fruit is bitter and inedible.

CULTIVARS Hundreds of cultivars have been selected over the centuries. Those that have fruited with some regularity in Florida include 'Ascolano,' a cultivar that bears heavy crops of large fruit with a small pit; 'Manzanillo,' a cultivar that bears reasonably well in north Florida and south Georgia; 'Barouni,' a Tunisian cultivar, tolerant of high temperatures, that bears a fruit of excellent quality; and 'Lucca,' a California cultivar that has been planted to some extent in the Southeast. Other promising selections include 'Mission,' 'Sevillano,' and 'Picholine.'

RELATIVES The family Oleaceae consists of about 25 to 40 genera and 500 to 600 species. It includes flowering ornamentals such as jasmine, *Jasminum* spp.; lilac, *Syringa* spp.; and forsythia, *Abeliophyllum* spp. Several olive-relatives are native to Florida, including the devilwood, *Osmanthus americanus,* of north and central Florida; the scrub olive, *Osmanthus megacarpa,* an endangered tree native to central Florida; the fringe tree, *Chionanthus virginicus,* native to much of the eastern United States; and several privets, *Forestiera* spp.

CLIMATE The olive prefers a mild, dry climate. High humidity has an adverse impact on production. The tree produces satisfactory vegetative growth but will not fruit in south Florida. Fruiting is rare in central Florida. Most cultivars appear to require between 200 and 400 hours of temperatures below 45° F. Fruiting is said to occur with some regularity in Jacksonville. The tree may be injured when the temperature falls below 10° F.

CULTIVATION The olive is a hardy, low-maintenance tree. It is drought-resistant. It will grow in rocky and nutrient-poor soils. It is tolerant of alkaline soils and moderate salinity. Frequent sprays of fungicide should be administered to prevent fruit drop caused by humid conditions. Few pests affect the olive in Florida. The tree is most often reproduced through hardwood cuttings.

HARVEST AND USE The fruit ripens in the fall or winter. The fresh fruit contains the glucoside oleuropein, and is exceedingly bitter. It is soaked in a lye solution to neutralize this compound. Do not use an aluminum pot to cure olives, as the lye solution will leach zinc from the pot. After soaking for 12 hours, the fruit is transferred to a fresh solution. It is then soaked for a week in several changes of cold water until any residual lye has been eliminated. The fruit can then be sundried, pickled in vinegar for long-term storage, or cured in oil.

This tree, at the Fruit and Spice Park in Homestead, shows that the olive will grow in all regions of the state.

For the most part, fruit production is limited to north Florida

Orange

SCIENTIFIC NAME: *Citrus sinensis*
FAMILY: Rutaceae
OTHER COMMON NAME: Naranja (Spanish)

Fruiting Calendar

JAN	FEB	MAR	APR	MAY	JUN	JUL	AUG	SEP	OCT	NOV	DEC

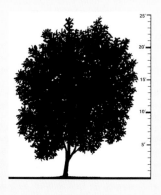

Characteristics

Overall Rating	★★★★★
Ease of Care	★★★★
Taste/Quality	★★★★★
Productivity	★★★★
Landscape Value	★★★★
Wind Tolerance	★★★
Salt Tolerance	★★
Drought Tolerance	★★★
Flood Tolerance	★★★
Cold Tolerance	★★★

From an economic perspective, the sweet orange is the most important species within the *Citrus* genus. The orange blossom is the state flower of Florida. The species is grown in countless yards across the state, despite the fact that the fruit is inexpensive and readily available from every market. The tree is beautiful, productive, and requires little maintenance. However, the spread of citrus canker and citrus greening presents a major threat to the future of the orange as a dooryard fruit crop.

Known Hazards

Oils in the fruit rind and leaves may cause dermatitis in sensitive individuals.

GEOGRAPHIC DISTRIBUTION The center of diversity for citrus extends from India east to the Malaysian archipelago and north to southern China. It is thought that the orange originated in India or southern China. The orange reached the Mediterranean prior to 1500 and was introduced into the Americas by the Spanish. Today, Florida leads the nation in production.

TREE DESCRIPTION The orange is a medium-size, densely foliated evergreen with a rounded, roughly symmetrical crown. It typically attains a height of about 25 feet with an equal spread, but may attain a greater size under ideal conditions. The deep green, somewhat leathery leaves are elliptic to ovate, and measure between 3 and 5 inches in length. The leaves may be entire or lightly toothed. The leaf petioles bear wings that range from indistinct to prominent. Thin gray bark sheaths the trunk and older growth.

The fragrant flowers, which measure from 1 to 1 1/2 inches in diameter, are borne on the current season's growth. The 5 petals are white, thick, and somewhat waxy. The stamens, between 20 and 25 in number, are topped with yellow anthers. Most cultivars are self-pollinating. Some cultivars are parthenocarpic—that is, they will set and mature seedless fruit without fertilization.

FRUIT CHARACTERISTICS As with other *Citrus* species, the fruit of the orange is a hesperidium—a modified berry. Globose to oblate in form, it typically measures between 2 1/2 and 3 1/2 inches in diameter. The skin is thin and aromatic and contains numerous oil glands. As with other forms of citrus, oils in the peel may cause dermatitis in some individuals. The inner rind is white, spongy, and bitter. The pulp is made up of 10 to 14 sections, divided by thin septa. These may be seedless or may contain up to 4 seeds. Each flesh segment is composed of hundreds of tiny juice vesicles. Sugar content is high.

CULTIVARS In Florida, cultivars are usually grouped according to the harvest season, as early, mid-season, or late.

'Navel' - This early-season variety produces large fruit of excellent quality. The fruit are easy to peel and are relished for eating out of hand. The 'Navel' orange is thought to have originated in Brazil in the early 1800s. Fruit ripens from early November to January. Yield is moderate but consistent.

'Valencia' - This is the premier late-season cultivar in Florida. It bears from March through May. 'Valencia' probably originated in Portugal. It was first planted in Florida in 1877. The medium fruit have few seeds and are of excellent quality for both eating out of hand and juicing. Yield is high and consistent.

'Pineapple' - 'Pineapple' is a mid-season cultivar of commercial import. It was selected in Florida around 1860. The medium-large fruit are of very-good flavor and texture, but can be somewhat seedy. Harvest season runs from December through early February.

'Hamlin' - This cultivar was selected in Florida in the late 1800s. It produces fruit from late October through January. The tree has displayed considerable cold tolerance. Fruit quality is very good. Yield is consistently high and fruit stores well on the tree.

'Queen' - This mid-season cultivar bears medium fruit that are of good quality but seedy. It was selected in Florida around 1900. 'Queen' is similar to 'Pineapple' with respect to fruit quality and bearing habits.

'Ambersweet' - 'Ambersweet' is an early-season variety that is actually the result of cross between a sweet orange and a tangerine-tangelo hybrid. It bears fruit of good flavor, which may contain numerous seeds. Harvest is from October to January.

'Parson Brown' - This early variety ripens immediately before 'Hamlin.' However, it is slightly less productive and the fruit, while of good quality, is seedy. Fruit ripens from November through January.

Other cultivars that are sometimes grown in Florida include early season varieties, 'Earligold,' 'Itaborai,' 'Trovita,' Vernia,' and 'Westin'; mid-season varieties, 'Gardner' 'Homosassa,' 'Jaffa,' 'Midsweet,' 'Sanford Mediterranean,' and 'Sunstar'; and late-season variety 'Pope Summer.'

RELATIVES The *Citrus* genus is the most economically important sector of the medium-size Rutaceae family. It includes the citron, *Citrus medica;* the grapefruit, *Citrus* x *paradisi;* the Kaffir lime, *Citrus hystrix;* the Key lime, *Citrus aurantifolia;* the lemon, *Citrus limon;* the pummelo, *Citrus maxima;* the sour orange, *Citrus aurantium;* and the mandarin, *Citrus reticulata.* Several members of the *Citrus* genus closely resemble the sweet orange. The tangelo, *Citrus* x *tangelo,* is the result of a cross between the mandarin and the grapefruit. It is a favorite dooryard fruit tree in south Florida. Unlike the orange, the fruit may have a slight neck, giving it a characteristic 'bell' shape. Fruit ripen over the winter. The two most notable cultivars are 'Minneola' and 'Orlando.' Both are hybrids of the

The calomondin, a relative of the orange, is cold tolerant. Unfortunately, the fruit is intensely acidic and is scarcely edible out of hand.

grapefruit and 'Dancy' tangerine. The tangor is a cross between the mandarin and the orange. The most important cultivar in Florida is 'Temple,' often casually but incorrectly referred to as the "Temple Orange." Cultural requirements of the tangelo and tangor are nearly identical to those of the orange.

The family Rutaceae also includes diverse noncitrus species. The limeberry, *Triphasia trifolia,* native to Indonesia, is a shrub that is occasionally grown in south Florida. Flowers are small, white, and fragrant. The small, bitter fruit is dull red. The wampee, *Clausena lansium,* native to Southeast Asia, has been grown in Florida since at least 1908. The small tree bears clusters of 1-inch fruit, with 5 longitudinal ridges. The wampee is subtropical in habit and may be killed or severely injured at temperatures below 25° F. The elephant apple, *Feronia limonia,* native to India, bears a 2- to 5-inch fruit with a hard rind. The pulp is brown, sticky, and aromatic, ranging in flavor from acidic to sweet. The curry leaf tree, *Murraya koenigii,* is a small, deciduous tree native to India. The aromatic leaf is an essential ingredient in many Indian dishes. It is becoming increasingly popular in Florida home gardens. Several Rutaceae species are native to Florida although none bear fruit of merit, with the possible exception of the sea torchwood, *Amyris elemifera,* which bears a purple drupe that is said to be edible. Noncitrus Rutaceae species covered within this book include the bael fruit, *Aegle marmelos;* kumquat, *Fortunella* spp.; and white sapote, *Casimiroa edulis.*

CLIMATE The orange is best adapted to areas of the state south of Ocala and, today, is primarily grown from the Orlando area southward. In times past,

oranges have been grown in north Florida and even in coastal areas of Georgia. For many years the citrus industry was based in central and north central Florida. Over the years, the main growing areas have been pushed southward by a series of freezes. Depending on tree health and the degree of hardening off, a mature tree may survive temperatures of 24° F or lower. A tree that is in an active state of growth can suffer serious damage at 26° F. Fruit may be killed or severely damaged by a temperature drop to between 28° and 26° F.

CULTIVATION The tree requires low to moderate maintenance. It is adapted to a wide range of soils and will grow on sand, loam, clay, limestone soil, and even on somewhat poorly drained soils, presuming that an appropriate rootstock is chosen. It is important to keep the area beneath the canopy free of competing weeds and grasses, especially when the tree is young. However, mulch is rarely used for this purpose, as it tends to spread fungal rots into the bark and trunk. A few backyard growers use mulch, but are careful to ensure that it is well removed from the area around the base of the trunk.

Irrigation is critical during periods of low rainfall. However, excessive irrigation during the latter phases of fruit development can negatively affect flavor. At the same time, failure to provide sufficient irrigation may lead to premature fruit drop. The grower should be aware that some fruit drop is normal and does not necessarily signal that the tree needs water.

The young tree should receive several light applications of balanced fertilizer, spaced throughout the year, to ensure a continuous supply of nutrients. Slow-release products allow for less frequent applications. Supplemental foliar applications of minor nutrients have been found to accelerate growth. To ensure steady production, a mature tree should receive between 1 and 2 pounds of nitrogen per year. A heavy application is usually administered around the time of harvest to promote bloom and new growth. The tree does not require pruning, except to remove crossing branches, deadwood, and water sprouts.

PESTS AND DISEASES Various pests attack the orange in Florida. Most insect pests are relatively easy to control and many insecticides are approved for citrus. The citrus leafminer is a common but fairly minor pest of orange and other dooryard citrus. It leaves a meandering trail of damage on the underside of emerging leaves, leading to cupping and deformation of foliage. Scale insects damage foliage and can lead to gradual decline of the tree. Several species of aphids attack emerging foliage.

The orange is moderately wind resistant. This 'Navel' orange, uprooted by Hurricane Jeanne, could not be salvaged.

Infestation by the citrus rust mite causes a bronzing of the fruit skin. The damage is primarily cosmetic and reduces fruit marketability. The citrus red mite feeds on the leaves and fruit. Other insect pests include root weevils, the citrus blackfly, various whiteflies, and the citrus mealybug. Nematodes cause root damage and tree decline.

Several serious diseases also attack the orange. Citrus canker affects the orange and other *Citrus* species. The 'Hamlin,' 'Navel,' and 'Pineapple' cultivars are particularly vulnerable. Citrus greening, another serious and uniformly lethal disease, represents a new threat to the orange in Florida. No effective treatment has been discovered. Greasy spot, a fungal disease, can result in severe defoliation. Symptoms include a yellow and brown spotting or mottling of leaves, which eventually become lesions with a dark, greasy appearance.

PROPAGATION Unlike many fruit species, the orange often comes true from seed. Freshly harvested seeds germinate readily, usually within a month. However, seedlings are slow to come into production. Virtually all trees raised in Florida are budded or grafted onto sour orange, citrange, or other rootstock. Chip budding and T-budding are the most common techniques.

HARVEST AND USE Fruit mature 9 to 16 months after bloom. In Florida, harvest takes place from late October through May. Yield typically ranges from 50 to 300 fruit per tree. Maturity should be determined through tasting, as orange peel color does not always correspond with physical maturity. The fruit can be stored on the tree for several weeks after it reaches physical maturity. The orange can also be refrigerated for up to 3 months.

The beauty of citrus serves to underscore its importance to Florida as a commercial and dooryard crop.

Orange

Papaya

SCIENTIFIC NAME: *Carica papaya*
FAMILY: Caricaceae
OTHER COMMON NAMES: Pawpaw, Melón Zapote or
  Fruta Bomba (Spanish)

Fruiting Calendar

JAN	FEB	MAR	APR	MAY	JUN	JUL	AUG	SEP	OCT	NOV	DEC

Characteristics

Overall Rating	★★
Ease of Care	★★
Taste/Quality	★★★
Productivity	★★★★
Landscape Value	★★★
Wind Tolerance	★
Salt Tolerance	★
Drought Tolerance	★★
Flood Tolerance	★
Cold Tolerance	★

The papaya is a major tropical fruit, widely cultivated throughout the lowland tropics. The fruit is delicious and is amenable to many uses. The plant is compact, precocious, and produces a steady crop over many months. Unfortunately, it is susceptible to damage from cold temperature, flooding, and wind. It is also susceptible to attack by several pests and life-shortening diseases. Consequently, raising the papaya in Florida entails some effort and risk. For the motivated home gardener, the rewards may outweigh the disadvantages.

Known Hazards

The milky latex exuded by the plant has been known to cause contact dermatitis. Some individuals are severely allergic to all parts of the plant, including the fruit and pollen.

GEOGRAPHIC DISTRIBUTION The papaya is native to southern Mexico and Central America. It has been widely distributed around the globe and is grown in nearly every tropical region.

PLANT DESCRIPTION The plant is a large herb that may grow to 20 feet. Leaves are deeply lobed, palmate, and measure from 10 inches to over 2 feet in diameter. The plant usually has a single stem although, if injured, it may put out several stems. The stem is tubular and hollow and is marked with persistent leaf scars. While the plant ordinarily has useful life of about 5 years, viral diseases have drastically reduced this period in Florida. Flowers may be female, bisexual, or male. Female flowers are born on short pedicels. They are white, thick, fleshy, and have 5 petals. Male flowers are smaller and are born on branched panicles.

FRUIT CHARACTERISTICS The fruit is melonlike and ranges in size from about 8 ounces to 12 pounds or more. The flesh may be yellow, orange, pink, or red. The flavor is sweet, aromatic, and musky. A hollow cavity, rimmed with hundreds of small dark seeds, runs through the center of the fruit.

CULTIVARS Varieties recommended for Florida, include 'Cariflora,' 'Red Lady,' and 'Yellow Lady.' Hawaiian varieties 'Solo,' 'Sunrise Solo,' and 'Sunset Solo,' which are found in the supermarket, grow readily from seed and produce small fruit of high quality. However, the plants usually succumb quickly to disease in Florida.

RELATIVES The Caricaceae family is small, composed of 5 genera and 30 to 50 species. The babaco, *Carica* x *heilbornii* var. *pentagona,* is cultivated at middle elevations of the Andes. The elongated fruit has white pulp and is not as sweet as a typical papaya. The chamburo, *Carica pubescens,* is native to the northern Andes. The small yellow fruit contains tart flesh. The mountain papaya, *Carica candamarcencis,* is native to the high Andes. The fruit is insipid and is cooked prior to consumption.

CLIMATE The plant may be damaged or killed by a brief temperature drop to 30° F. However, the papaya can be grown as a single-season crop in north Florida if given a head start indoors.

CULTIVATION The papaya has fairly high maintenance requirements. Brief flooding or persistently waterlogged soil will kill the plant. It is easily damaged by high wind. It requires regular irrigation during warm weather. It benefits from frequent applications of a balanced fertilizer. New plants should be started on a regular basis to replace those that are lost to age and disease. When setting out container-grown plants, great care must be taken not to disturb the sensitive root system.

The most serious insect pest is the wasplike papaya fruit fly. The female has a long ovipositor that it uses to pierce the skin of the fruit and lay eggs. Upon hatching, the maggots consume the interior of the fruit. Where this insect is present, the grower must bag the fruit. Other insect pests include thrips, white flies, mites, and the papaya webworm. Nematodes are a serious problem in sandy soil. The papaya ring spot virus causes dark green rings on the fruit. Leaves become stringy and mottled. Infected plants should be destroyed. Most casual growers propagate the plant by seed.

HARVEST AND USE Fruit production peaks over the summer and fall. The fruit is mature when the skin color begins to change. Once harvested, it should be allowed to ripen at room temperature. The fruit can be halved and eaten plain or used in fruit salads and chutneys.

The bud of the female flower of the papaya.

Fruit come ripe one at a time, providing a near-constant supply.

Passionfruit

SCIENTIFIC NAME: *Passiflora edulis*
FAMILY: Passifloraceae
OTHER COMMON NAME: Granadilla or Ceibey (Spanish)

Fruiting Calendar

JAN	FEB	MAR	APR	MAY	JUN	JUL	AUG	SEP	OCT	NOV	DEC

Characteristics

Overall Rating	★★★
Ease of Care	★★
Taste/Quality	★★★★
Productivity	★★★
Landscape Value	★★★
Wind Tolerance	★★
Salt Tolerance	★
Drought Tolerance	★★
Flood Tolerance	★★
Cold Tolerance	★★

Those familiar with the passionfruit consider it among the world's finest vine-borne fruit, rivaling the grape in overall appeal. While the pulp is sparse and is intermingled with seeds, a small quantity is adequate for most purposes. The plant itself is ornamental. The flowers are exquisite. This species needs a trellis or other support and requires some maintenance. Nevertheless, it is one of the most rewarding dooryard species that can be grown in south Florida.

Known Hazards

Some species have invasive-exotic tendencies.

GEOGRAPHIC DISTRIBUTION The passionfruit is native to South America. It comes in two forms. The purple passionfruit originated in the area stretching from southern Brazil to northern Argentina. The yellow passionfruit may have originated in the Amazon region of Brazil or may be a mutation of the purple passionfruit. The passionfruit has been distributed to many subtropical and tropical regions. A few small commercial operations exist in Florida. Several *Passiflora* species have escaped cultivation and have displayed invasive tendencies.

PLANT DESCRIPTION The passionfruit is a woody, perennial vine. It is fast growing and is a vigorous climber. It fastens itself to trees, fences, trellises, or other structures with tendrils that emerge from the leaf axils. The vine may reach a length of 30 feet or more. The root system is shallow but spreading. Leaves are alternate, lobed, with toothed margins. They measure between 3 and 8 inches in length. Plants started from seed will usually bear within a year. The vine is short lived, and generally declines after 3 to 5 years.

FLOWERING AND POLLINATION Flower buds are produced at the leaf nodes on the current season's growth. The flower is highly ornamental. It measures between 2 and 5 inches in diameter. The ovary and 3-branched style protrude above 5 stamens with large anthers. These center organs are surrounded by a corona composed of threadlike filaments, which is, in turn, surrounded by 5 petals and 5 sepals.

The flower of the yellow passionfruit, which typically opens during the afternoon, is self-sterile. Visiting insects, especially carpenter bees, act as pollinating agents. The purple passion fruit is self-pollinating. Hand pollination is sometimes practiced to increase fruit set.

FRUIT CHARACTERISTICS The fruit, botanically classified as a berry, may be round or ovoid. It measures from 1 1/2 to 3 inches in diameter. The skin is glossy and smooth. Depending on the form or cultivar, the color may change from green to purple, red, pink, or yellow, as the fruit ripens. The rind softens and becomes leathery as the fruit matures. A white-walled membrane—the endocarp—encloses a central cavity that contains the golden-yellow pulp, actually the seed arils. The fruit contains between 100 and 300 small, dark seeds. The flavor is highly distinctive. It is rich, aromatic, sweet, and tangy.

FORMS AND CULTIVARS The purple passionfruit, although commonly planted in Florida, grows and fruits poorly in the humid tropical lowlands. The yellow passionfruit is more vigorous and more pro-

A chain-link fence is not the ideal trellis, but it will generally suffice.

ductive. It bears fruit that are slightly larger and more acidic than those produced by the purple passionfruit. The yellow passionfruit is widely cultivated in the lowland tropics. Numerous hybrids exist between the purple and the yellow passionfruit. In Florida, the selection of passionfruit cultivars is limited. Many strains are grown from seed. Seeds can be ordered by mail or over the Internet from several suppliers. Notable cultivars include 'Black Beauty,' 'Black Knight,' 'Edgehill,' 'Frederick,' 'Golden Giant,' 'Kahuna,' 'Lacy,' 'Noel's Special,' 'Panama Gold,' 'Panama Red,' 'Possum,' 'Pratt Hybrid,' 'Purple Giant,' 'Red Giant,' and 'Red Rover.'

RELATIVES The Passiflora or Passionflower family, within the order Violales, is a relatively small family composed of about 15 genera and 400 to 500 species—primarily vines. At least 60 species produce edible fruit. Several species have been grown for their fruit in Florida.

The giant granadilla, *Passiflora quadrangularis,* is native to the lowlands of tropical America, although its precise origin is uncertain. Flowers are large, measuring up to 5 inches in diameter, with white petals and purple filaments. The melonlike fruit is the largest within the *Passiflora* genus, measuring up to 10 inches in length and 6 inches in diameter. Both the flesh and the arils are edible. While the pulp is sweet and aromatic, it does not equal that produced by the common passionfruit or the water lemon. The giant granadilla is occasionally grown in south Florida.

The maypop, *Passiflora incarnata,* is a hardy native passionfruit, growing throughout most of the eastern United States. The name reflects the fact that the plant dies back over the winter and sprouts vigorously upon the return of warm weather. Unlike the fruit produced by some native plants, the fruit of the maypop is not simply edible, but is actually worth

Many members of the Passiflora family produce stunning flowers.
On the left is the variety 'Incense' and on the right is the giant granadilla.

eating. The fruit is ovoid, measuring 1 1/2 to 2 inches in diameter. It is highly recommended for planting in central and north Florida. Several *Passiflora* species, in addition to the maypop, are native to Florida, including the yellow maypop, *Passiflora lutea;* the corky-stem passionflower, *Passiflora suberosa;* the goatsfoot, *Passiflora sexflora;* the pineland passionvine, *Passiflora pallens;* and the white-flower passionflower, *Passiflora multiflora.*

The water lemon or yellow granadilla, *Passiflora laurifolia,* is native to northern South America, the Amazon region, and the West Indies. Although tropical in habit, it is reasonably well suited to cultivation in south Florida. The yellow fruit, which outwardly resembles a lemon in shape and size, contains sweet, high-quality pulp.

The melonlike fruit of the giant granadilla, shown here nestled in a tangle of vines.

Other fruiting members of the *Passiflora* genus native to South America include the banana pas-

sionfruit, *Passiflora mollissima,* which is poorly adapted to Florida, but which produces a yellow, oblong fruit of excellent quality; the bell apple, *Passiflora nitida,* which produces a 3-inch fruit with a brittle rind and gray pulp of good quality; the puru-puru, *Passiflora pinnatistipula,* which produces a round fruit with grayish-yellow pulp of good quality; the curuba, *Passiflora mixta,* which produces oblong fruit that varies in size and quality; the fetid passionflower, *Passiflora foetida,* which is considered an invasive exotic in Florida, but which bears fruit of good quality; the fragrant granadilla, *Passiflora alata,* which bears 7-inch-long fruit with orange pulp of excellent quality; the granadilla de Quijos, *Passiflora popenovii,* which bears 3-inch fruit of good quality; the grape-leaved passionflower, *Passiflora vitifolia,* which bears acidic fruit of good quality; the machimbi, *Passiflora tiliaefolia,* which produces 3-inch fruit with gray pulp of superb quality; the parcha de monte, *Passiflora cincinnata,* which bears medium-large fruit of fair quality; the red granadilla, *Passiflora coccinea,* which bears medium fruit of good quality; the sweet calabash, *Passiflora maliformis,* which bears a medium fruit with a shell-like rind; the sweet granadilla, *Passiflora ligularis,* which bears a fruit of very good quality; and the taguatagua, *Passiflora serrato,* which produces medium fruit of good quality, said to resemble the guava in flavor.

CLIMATE The purple form of the passionfruit is subtropical in habit. It suffers leaf damage at about 30° F, but may withstand a temperature drop to 25° F. The yellow form is more sensitive to cold and may be killed at 27° F. The yellow form may be slightly better suited to growth in south Florida than the purple form, as it is more tolerant of heat and humidity. The passionfruit is grown as far north as Cape Canaveral on the east coast and Pinellas County on the west coast.

CULTIVATION The passionfruit vine has relatively high maintenance requirements. It is not particular as to soil type but good drainage is essential. The plant is particularly susceptible to root rot. It prefers a soil pH of 6.0 to 7.0. It also benefits from a heavy application of mulch, which serves to reduce nematode populations and to retain surface moisture.

The plant should receive adequate sun and may be planted in full sun. However, light, mid-day shade may be beneficial during periods of extreme summer heat. Thick shade will cause a significant decrease in production.

The species prefers an even distribution of rainfall throughout the year. Consequently, in Florida, irrigation is important during the dry season, during flowering, and during periods when the vine holds a heavy load of fruit. Because the root system is shallow, the plant may begin to wilt or drop fruit if the first few inches of soil are permitted to dry out. Water stress causes a decline in production, as the plant does not flower under dry conditions.

Regular applications of fertilizer are needed to sustain the plant's rapid growth rate. High-nitrogen fertilizers may be used to stimulate growth until the plant reaches bearing age. After that, a balanced fertilizer should be applied. A mature plant should receive 3 or 4 applications over the warmer months.

The vine may be trained to grow on a fence, trellis, wall, or arbor. The main stem should be loosely tied to a pole, with all side growth removed until the plant reaches the top of the structure. At that point, the main leader is pinched to encourage lateral growth. Some homeowners avoid the need for a trellis by allowing the vines to climb trees, but squirrels tend to find the fruit under such circum-stances. The vine should be severely cut back in the early spring to encourage new growth. Promoting new growth is critical to production, since the fruit are borne on the current season's growth. Laterals should be cut back by 1/3 annually, and about 1/3 of the laterals should be removed. Any shoots trailing on the ground should be removed. The vine must be pruned and kept in bounds as it will otherwise form a mass of tangled foliage.

PESTS AND DISEASES In Florida's sandy soil, the passionfruit is bothered by nematodes, which can reduce the productive lifespan of the plant. The yellow form appears to be less susceptible to damage than the purple form. Squirrels raid the crop and, if left unchecked, they are capable of completely stifling production.

PROPAGATION In Florida, the vine is usually grown from seed. Seeds germinate readily, within 2 or 3 weeks after they are sown. They can be stored under dry conditions for several months without a significant reduction in the rate of germination. This species has also been propagated through cuttings of mature wood.

HARVEST AND USE Fruit matures between 2 and 3 months after bloom. It can be picked when fully colored or can be gathered after it has fallen to the ground. The prolonged harvest season peaks during in the late summer and early fall. The fruit will store for an extended period. It is at its best after the skin shrivels slightly. The rind and seeds typically make up more than 60 percent of the weight of the fruit. However, the flavor is strong and a little pulp goes a long way. The fruit can be eaten out of hand or used as a flavoring for ice cream, sherbet, beverages, and sauces.

Pawpaw

SCIENTIFIC NAME: *Asimina triloba*
FAMILY: Annonacea
OTHER COMMON NAME: Hoosier Banana

Fruiting Calendar

JAN	FEB	MAR	APR	MAY	JUN	JUL	AUG	SEP	OCT	NOV	DEC

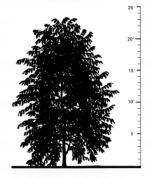

Characteristics

Overall Rating	★★★★
Ease of Care	★★
Taste/Quality	★★★★
Productivity	★★
Landscape Value	★★★
Wind Tolerance	★★★
Salt Tolerance	★
Drought Tolerance	★★
Flood Tolerance	★★★★
Cold Tolerance	★★★★★

This temperate member of the Annona family is the largest tree-borne fruit native to the continental United States. The pawpaw is a fascinating species, worthy of improvement. Selected cultivars produce fruit of outstanding flavor. It is unfortunate that agricultural interests within the United States have been slow to recognize and promote this species. Enterprising growers in other nations are sure to adopt this unique fruit and market it to the rest of the world.

Known Hazards

The contents of the seed are toxic. A few individuals are allergic and develop dermatitis after touching the fruit.

172

GEOGRAPHIC DISTRIBUTION The common pawpaw ranges from New York, west to Michigan, and south to the Gulf coast. It grows wild in the panhandle. A specimen is present on the grounds of the governor's mansion in Tallahassee.

TREE DESCRIPTION The pawpaw is a small, deciduous tree, commonly attaining a height of 15 to 20 feet. Leaves, which measure from 6 to 10 inches in length, are dark green and drooping. The tree sometimes reproduces by root suckers, forming multiple trunks. The pawpaw is an under-story plant and requires partial shade, particularly when young. The purple or maroon flower, which appears in the spring, has 6 petals in 2 whorls of 3. Each flower is capable of producing more than one fruit. Two or more cultivars should be planted to ensure pollination.

FRUIT CHARACTERISTICS The fruit superficially resembles a green-skinned mango. It may measure up to 8 inches in length and may weigh over a pound. More typically, it weighs between 6 and 10 ounces. The skin is smooth, thin, and edible. The pulp is custardlike, ranging from light yellow to orange. In flavor, it resembles a blend of banana, mango, and papaya. Fruit from a single flower may form a tightly packed cluster. Seeds, usually numbering between 4 and 10, are positioned in a row toward the center of the flesh.

CULTIVARS At least 50 cultivars have been selected. 'Overleese,' 'Collins,' 'Mango,' 'Blue Ridge,' and 'Duckworth' are said to perform well in north Florida. 'Sunflower,' 'Wells,' 'Taytoo,' 'Prolific,' and 'Shenandoah' are well-regarded northern cultivars that have not been sufficiently tested in Florida.

RELATIVES The Annonaceae family contains many fruiting species. Other Annonas covered within this book are the atemoya, *Annona cherimola* x *Annona squamosa;* custard apple, *Annona reticulata;* and sugar apple, *Annona squamosa.* Various tropical Annonas are discussed within the subsection pertaining to relatives of the atemoya. The genus *Asimina* contains several species native to Florida, which bear small, edible fruit of variable quality. The small-flower pawpaw, *Asimina parviflora,* is a tall shrub that grows as far south as Highland County. The flag pawpaw, *Asimina obovata,* is a rare plant native to central Florida. The endangered four-petal pawpaw, *Asimina tetramera,* grows in the coastal scrublands of Martin County and Palm Beach County. The netted pawpaw, *Asimina reticulata,* is native to south Florida. Other native pawpaws of uncertain fruiting characteristics include the endangered beautiful pawpaw, *Deeringothamnus pulchellus;* the dwarf pawpaw, *Asimina pygmea;* the endandered Rugel's

pawpaw, *Deeringothamnus rugelii;* the slimleaf pawpaw, *Asimina angustifolia;* and the woolly pawpaw, *Asimina incana.*

The four-petal pawpaw is sensationally beautiful but, sadly, is teetering on the brink of extinction. Most flowers have three petals, despite the name.

CLIMATE The pawpaw can withstand temperatures as low as -20° F. Most varieties require 300 to 400 hours of chilling before they will set fruit.

CULTIVATION The pawpaw has moderate maintenance requirements. It should receive periodic irrigation during periods of active growth. The tree benefits from twice-annual fertilization: once after breaking dormancy in the spring, once in early summer. The young tree must be planted in shade or must be protected with shade cloth until it reaches about 4 feet in height. A mature tree can endure direct sun. The pawpaw has few pests or diseases. Squirrels, opossums, and raccoons sometimes raid the tree.

The pawpaw will grow readily from seed. However, seeds must be kept moist and must be stored at low temperature for several months prior to planting. Cleft grafting, whip-and-tongue grafting, and chip budding have been successfully employed as methods of vegetative propagation.

HARVEST AND USE The pawpaw usually ripens between August and November. A productive cultivar may yield 60 pounds of fruit. Maturity is indicated by a change in skin color, from green to light green or yellow, and easy separation from the branch. The pawpaw has a short shelf life and will spoil in less than a week at room temperature. It is generally chilled and cut lengthwise, with the flesh spooned from each half. The pulp can be used in custards or as a flavoring for ice cream.

The sweet, custardlike flesh of the pawpaw is a delicious early fall treat.

Peach

SCIENTIFIC NAME: *Prunus persica*
FAMILY: Rosaceae
OTHER COMMON NAME: Melocotón (Spanish)

Fruiting Calendar

JAN	FEB	MAR	APR	MAY	JUN	JUL	AUG	SEP	OCT	NOV	DEC

Characteristics

Overall Rating	★★
Ease of Care	★
Taste/Quality	★★★★
Productivity	★★★
Landscape Value	★★★
Wind Tolerance	★★★
Salt Tolerance	★
Drought Tolerance	★★★
Flood Tolerance	★★★
Cold Tolerance	★★★★★

Although most peach cultivars require significant winter chill, several will grow in north Florida. A few low-chill cultivars will succeed in central and southern portions of the peninsula. The tree demands considerable maintenance, as it is not well adapted to the humid subtropics. Many pests and diseases bedevil the peach in Florida. With that said, the peach is capable of producing a delicious dooryard crop for the determined grower.

Known Hazards

The interior of the pit is poisonous. Twigs and leaves are toxic.

GEOGRAPHIC DISTRIBUTION The peach is native to China. Peach cultivation spread westward along the silk trade routes, eventually reaching the Mediterranean. The Spanish are credited with introducing the peach to the Western hemisphere.

TREE DESCRIPTION The peach is a small, deciduous tree that can attain a height of 20 to 25 feet. The crown is spreading and irregular. Leaves, measuring from 4 to 7 inches in length, are alternate, lanceolate, with serrate margins and pointed tips. The tree has a short productive life and may begin to decline after about 20 years. It has fair to good landscape value. Flowers are pink or white and have 5 petals. The flowers are borne on bare branches from February through March depending on the cultivar and location.

The peach blossom, like the flowers of many members of the Rosaceae family, is highly decorative.

FRUIT CHARACTERISTICS The fruit is a large drupe, measuring 2 to 3 1/2 inches in diameter. The skin is thin and pubescent. Skin color ranges from green to yellow with a reddish blush. The sweet flesh is juicy, tender, and slightly fibrous. Flesh color ranges from white to yellow orange. The medium-size stone, a bony endocarp, is marked with deep ridges. It contains a single ovate seed that is highly toxic.

CULTIVARS More than 1,000 varieties have been selected worldwide. Since 1952, the University of Florida has been breeding and selecting stone fruit capable of producing fruit under subtropical conditions. Germplasm from this program has been distributed around the world and includes many of the cultivars discussed below. Some peach cultivars suitable for growth in north Florida include 'Gulfprince,' 'Floracrest,' 'Floradawn,' 'Floraking,' 'Junegold,' 'Maygold,' 'Suwannee,' and 'UF 2000.' Varieties for central Florida include 'Floraprince,' 'Tropic Prince,' 'Flordaglo,' 'Tropic Snow,' 'Tropic Sweet,' 'Tropic Beauty,' and 'UF Gold.' Peaches suitable for south Florida include 'Red Ceylon' and 'Flordagrande.'

RELATIVES The peach is a member of the diverse and fruitful Rosaceae family. Members described within this book include the apple, *Malus domestica;* blackberry, *Rubus* spp.; capulin, *Prunus salicifolia;* chick-

asaw plum, *Prunus angustifolia;* loquat, *Eriobotrya japonica;* mayhaw, *Crataegus* spp.; pear, *Pyrus* spp.; and strawberry, *Fragaria ananassa.* The peach belongs to the Prunoideae or "stone fruit" subfamily, which includes cherries and plums.

CLIMATE The peach is a crop of temperate climates, and the tree is impervious to low temperatures in Florida. Nevertheless, blossoms may be damaged by late spring frosts in north Florida.

CULTIVATION The peach has significant maintenance requirements. It prefers slightly acidic soil with an open texture. Proper drainage is vital. Three annual applications of a 10–3–10 fertilizer, or some similar mixture, are recommended. Irrigation is important during periods of fruit development.

The tree requires considerable pruning. Laterals and suckers are removed to a height of 2 feet to form a strong central leader. Three or 4 vigorous shoots should be selected to form the major branches. A tree of bearing age should be pruned while it is dormant. Thinning of immature fruit is often required, since excessive fruit density reduces fruit size. The goal is to leave 1 fruit per 6 inches of branch length.

The peach has a number of serious insect pests in Florida, including the Caribbean fruit fly, white peach scale, peach curculio, various borers, and stink buds. Root knot nematodes damage the roots of many cultivars, and may require the use of resistant rootstocks such as 'Flordaguard,' 'Nemaguard,' and 'Okinawa.' Diseases include rust, scab, brown rot, and phony peach. The peach is often propagated through T-budding.

HARVEST AND USE In Florida, the peach ripens from late April to July. The fruit is picked when it is still relatively firm. Maturity is adduced by a change in the ground color of the skin from green to greenish-yellow. The amount of blush is not a reliable indicator of fruit maturity. The peach is delicious when consumed fresh and can also be used in pies, cobblers, preserves, ice cream, and milkshakes.

The 'Ceylon' peach will grow and fruit in south Florida. However, the Caribbean fruit fly often ruins the fruit.

Pear

SCIENTIFIC NAME: *Pyrus* spp.
FAMILY: Rosaceae
OTHER COMMON NAME: Pera (Spanish)

Fruiting Calendar

JAN	FEB	MAR	APR	MAY	JUN	JUL	AUG	SEP	OCT	NOV	DEC

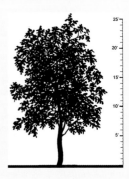

The pear is a temperate pome fruit. Like its close cousin, the apple, and its distant cousin, the peach, the pear is not especially well adapted to Florida. Many varieties require exposure to at least 800 chilling hours before they will produce fruit. However, a few cultivars are reasonably productive in north Florida. Where it succeeds, the pear is valued both for its beautiful springtime flowers and for its luscious fruit.

Characteristics

Overall Rating	★★★
Ease of Care	★★
Taste/Quality	★★★
Productivity	★★★
Landscape Value	★★★
Wind Tolerance	★★★
Salt Tolerance	★
Drought Tolerance	★★★
Flood Tolerance	★★★
Cold Tolerance	★★★★★

Known Hazards

The seeds contain variable toxins.

GEOGRAPHIC DISTRIBUTION The pear is thought to be indigenous to western Asia, although its progenitors may have been distributed throughout Asia and Europe. Several important pear cultivars were introduced into the United States from Europe in the 1700s and 1800s.

TREE DESCRIPTION The pear is a small to medium deciduous tree. Generally upright in form, it attains a height of about 25 or 30 feet. Leaves are alternate, simple, ovate, attaining a length of 2 to 4 inches. The margin is finely serrated. Clusters of white flowers, measuring from 1/2 to 1 inch in diameter, appear in the spring at about the time of bud break. Most cultivars require cross-pollination for adequate fruit set. The tree is long lived and makes an attractive landscape specimen.

FRUIT CHARACTERISTICS The fruit is a large pome, measuring between 2 and 4 1/2 inches in diameter. The true 'fruit' is limited to the 5 carpels that make up the core. The edible portion is actually formed from the flower receptacle. The shape ranges from pyriform to globose. The texture ranges from grainy to smooth and melting.

VARIETIES AND CULTIVARS Within the United States, three groups of pears are grown. The European pear, *Pyrus communis,* incorporates the great bulk of varieties commonly found on supermarket shelves, such as 'Anjou,' 'Bartlett,' and 'Bosc.' Most have sweet, melting flesh of excellent quality. The Asian pear, *Pyrus pyrifolia,* is more applelike in form, and has crisp flesh of very good quality. Oriental hybrid pears vary in quality but can equal European types in flavor. Pears suitable for growth in north Florida include 'Flordahome,' a productive cultivar, which requires between 200 and 350 chilling hours and which bears a medium-size fruit of fine texture; 'Hood,' a cultivar with a chilling requirement of about 350 hours, which bears a large fruit with yellow-green skin and fine texture; 'Le Conte,' a North Carolina cultivar with a chilling requirement of about 450 chilling hours, which bears a bell-shaped fruit of excellent quality; 'Pineapple,' a productive cultivar with a chilling requirement of about 350 or 400 hours; and 'Shinseiki,' an Asian cultivar with a chilling requirement of about 400 hours.

RELATIVES Other members of the Rosaceae family profiled within this book include the apple, *Malus domestica;* blackberry, *Rubus* spp.; capulin, *Prunus salicifolia;* chickasaw plum, *Prunus angustifolia;* loquat, *Eriobotrya japonica;* mayhaw, *Crataegus* spp.; peach, *Prunus persica;* and strawberry, *Fragaria ananassa.*

CLIMATE The pear is not affected by winter cold in Florida and the primary problem is lack of winter chill. Late frosts occasionally interfere with flowering. Even those varieties with very low chilling requirements do not fruit reliably in areas south of Ocala.

CULTIVATION The pear requires moderate cultural attention, although many specimens on old homesteads have lived for decades without any care. It prefers deep, well-drained soil, but can sometimes be grown on soils with less-than-optimal drainage. The tree should be pruned to a central leader system, in the same manner as the apple. About 2 months after bloom, the fruit should be thinned to 1 per spur. Several fungal diseases attack the pear, including rust and scab. Fire blight can be a severe problem and must be controlled through judicious pruning and the removal and destruction of infected limbs. The pear is most often propagated through chip or T-budding.

HARVEST AND USE While the Asian pear may be tree ripened, the European pear may develop gritty flesh if it remains on the tree until ripe. The European pear should be picked hard, once it has attained full size. It should then be allowed to ripen at room temperature. The fruit is ready to eat when the skin yields to gentle pressure. Ripe fruit can be stored for up to a week in the refrigerator and is not subject to chilling injury.

The pear makes a valuable addition to the home garden in north Florida.

Pecan

SCIENTIFIC NAME: *Carya illinoensis*
FAMILY: Juglandaceae
OTHER COMMON NAMES: Nogal Morado or Nuez
    Encarcelada (Spanish)

Fruiting Calendar

JAN	FEB	MAR	APR	MAY	JUN	JUL	AUG	SEP	OCT	NOV	DEC

Characteristics

Overall Rating	★★★★
Ease of Care	★★★
Taste/Quality	★★★★
Productivity	★★★★
Landscape Value	★★★
Wind Tolerance	★★★★
Salt Tolerance	★
Drought Tolerance	★★★★
Flood Tolerance	★★★
Cold Tolerance	★★★★★

The pecan is the most valuable nut crop of the southeastern United States. Although primarily a species for north Florida, it may also produce an acceptable crop in central Florida. The tree is beautiful, productive, and is not harmed by cold temperatures or summer heat. However, it is also large and somewhat slow to come into bearing. For those with sufficient property, the pecan makes a superb addition to the home landscape.

Known Hazards

None

GEOGRAPHIC DISTRIBUTION The pecan is native to the south central United States. It is found throughout the Mississippi bottomlands, north to Illinois. The pecan is widely planted in Georgia and has some commercial import in north Florida.

TREE DESCRIPTION The pecan is a large, upright, deciduous tree. It may eventually attain a height of 80 feet. Leaves are alternate, pinnately compound. The leaflets, numbering between 7 and 17, each measure from 4 to 6 inches in length. The leaflets are ovate-lanceolate and have a serrated margin. The tree is long lived. With its straight trunk, stately form, and feathery foliage, the pecan makes an attractive landscape specimen. However, its vertical proportions can dwarf the typical suburban home. The pecan should not be planted close to a dwelling, as the tree constantly sheds branches, sap, twigs, leaves, husks, and other litter. The pecan may require the presence of a pollinator to set a significant crop.

FRUIT CHARACTERISTICS The fruit is an elliptical nut enclosed in a thin husk. The husk is divided into 4 longitudinal segments, which split at maturity. The shell is brown, smooth and ovate. The kernel is divided into 2 hemispheres. In taste and overall appearance, the kernel resembles that of the walnut. However, it is sweeter, lighter, and more refined.

CULTIVARS Hundreds of varieties have been selected over time. The best cultivars for dooryard planting in Florida include 'Elliot,' a Florida cultivar that is resistant to scab and produces a small nut of exceptional quality; 'Moreland,' a Louisiana cultivar that produces heavy crops without cross-pollination; 'Curtis,' a Florida selection with good disease resistance, that has low chilling requirements and is productive in central Florida; 'Cape Fear,' a North Carolina cultivar that produces a nut of good quality; 'Stuart,' an old Mississippi cultivar that has performed consistently over time; 'Sumner,' a Georgia cultivar that is productive in central Florida; and 'Pawnee' a Texas cultivar that bears a large, high quality nut. Other worthwhile cultivars include 'Caddo,' 'Desirable,' 'Owens,' 'Gloria Grande,' 'Jenkins,' and 'Syrup Hill.'

RELATIVES The pecan is a member of the Walnut family, Juglandaceae. The hickory or *Carya* genus contains between 20 and 25 species. While at least 6 hickory species are native to Florida, none produce a nut of great value. The black walnut, *Juglans nigra*, is native to the panhandle and much of the eastern United States. The kernel is rich and flavorful, but the nut is exceedingly difficult to open.

CLIMATE The pecan is at home in warm-temperate climates. It can withstand temperatures as low as -15° F. The tree has some chilling requirements, but produces consistent crops as far south as Orlando. A tree on Merritt Island is said to bear with some regularity.

CULTIVATION The tree requires moderate maintenance. It prefers well-drained soil but can tolerate brief flooding. Weeds should be controlled around the base of the young tree. It should be pruned to a strong central leader. Adequate soil moisture is required from June through August to ensure that the nuts attain full size. A balanced fertilizer should be applied in March. The pecan has few serious pests. Squirrels sometimes consume the nuts and webworms sometimes damage the foliage. Scab is the most serious disease in Florida. The fungus attacks new leaves and nuts, causing them to drop before they fully develop. The best defense is to plant scab-resistant cultivars. Other diseases include blotch, powdery mildew, and rosette. Grafting is the preferred method of propagtion.

HARVEST AND USE The tree requires about 10 years before it will produce a significant crop. A mature tree may yield several hundred pounds on an annual basis. The nut usually falls when ripe. It may be consumed out of hand or may be used in pies, toppings, puddings, pralines, and confectionaries. The shelf life of the nut at room temperature is only about 3 months. It will store for extended periods if refrigerated or frozen.

The pecan grove provides a shady retreat during the hot north Florida summer.

Persimmon

SCIENTIFIC NAME: *Diospyros kaki*
FAMILY: Ebenaceae
OTHER COMMON NAMES: Oriental Persimmon, Caqui
  (Spanish)

Fruiting Calendar

JAN	FEB	MAR	APR	MAY	JUN	JUL	AUG	SEP	OCT	NOV	DEC

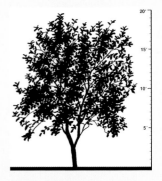

Characteristics

Overall Rating	★★★★
Ease of Care	★★★
Taste/Quality	★★★★
Productivity	★★★★
Landscape Value	★★★
Wind Tolerance	★★★
Salt Tolerance	★★
Drought Tolerance	★★★★
Flood Tolerance	★★★
Cold Tolerance	★★★★★

The persimmon produces a bountiful dooryard crop in Florida. It is most at home in north Florida, although it succeeds from the coldest parts of the panhandle to the near-tropical Keys. Because of its import in China and other populous nations, the persimmon is one of the world's most frequently consumed fruits. Fruit quality ranges from very good to superb. The tree demands little cultural attention and is an excellent choice for the home garden in Florida.

Known Hazards

None

GEOGRAPHIC DISTRIBUTION The oriental persimmon is indigenous to China, where it has been grown for thousands of years. It was first introduced into the United States in the late 1800s and was widely grown in Florida by the 1930s. Today, commercial orchards exist in the vicinity of Gainesville and Tallahassee.

TREE DESCRIPTION The persimmon is a small, deciduous tree. It may reach 25 feet in height but is usually smaller. Growth is slow to moderate. Leaves are alternate and ovate. They typically measure between 4 and 8 inches in length. Cream-colored flowers form in the leaf axils shortly after leaves emerge in the spring. Most named varieties bear only female flowers and produce seedless fruit without a pollinator. If male flowers are present, some varieties will produce seeded fruit.

FRUIT CHARACTERISTICS The fruit may weigh from a few ounces to nearly a pound, and may range from 1 1/2 to 4 inches in diameter. The shape varies with the cultivar. The color may be yellow, orange, or bright red. The flavor is mild, sweet, and pleasant. The texture is fine and smooth.

CULTIVARS Hundreds of cultivars exist in Japan and China. They fall within two broad categories: astringent and nonastringent. Nonastringent types are edible before they soften. Astringent persimmons only achieve full flavor and sweetness when their flesh turns gelatinous. Cultivars recommended for Florida include 'Fuyugaki,' a nonastringent cultivar, productive in north and central Florida, which bears orange fruit of good flavor; 'Saijo' an astringent persimmon, which bears yellow-orange fruit of exceptional quality; 'Triumph,' a high-quality astringent persimmon, which is productive throughout the state, including south Florida; 'Jiro,' a nonastringent cultivar that bears a lobed fruit of good quality on a compact tree; 'Tanenashi,' an astringent cultivar, which is productive throughout the state; 'Izu,' a nonastringent cultivar recommended for north and central Florida; and 'Hachiya,' an astringent cultivar, recommended for north and central Florida. Other cultivars that have potential in Florida include 'Eureka,' 'Sheng,' 'Great Wall,' 'MatsumotoWase Fuyu,' 'Midia,' 'Giombo' 'IchikikeiJiro,' and 'Suruga.'

RELATIVES The persimmon is a member of the Ebanaceae or Ebony family. Other notable fruiting species are the black sapote, *Diospyros dignia,* and mabolo, *Diospyros blancoi,* both discussed within these pages. The American persimmon, *Diospyros virginiana,* is a medium-size tree native to the eastern United States, including Florida. Those who favor native flora should consider planting this species. The fruit is very astringent until it softens. However, the fruit quality of some cultivars, including 'Early Golden,' 'Ennis,' 'John Rick,' 'Meador,' and 'Yates,' approaches that of the oriental persimmon. The Texas persimmon, *Diospyros texana,* is a shrublike tree ranging from Texas to Florida, and north to Connecticut. The seedy, dark purple fruit provides fodder for wildlife.

CLIMATE The persimmon can tolerate the most severe winter conditions likely to occur in Florida, with some cultivars surviving temperatures of -10° F. Some varieties have moderate chilling requirements.

CULTIVATION The tree is undemanding and is relatively free of problems. It can tolerate a wide range of soil, but prefers well-drained locations. It is fairly drought tolerant. The tree benefits from 2 light applications of a balanced fertilizer, the first just before leaf break in late winter, and the second in late spring or early summer. The persimmon should be pruned when young to establish a vase-shaped framework of limbs. Mealybugs, scale insects, and thrips sometimes inflict minor damage. Birds and various mammals, especially opossums, raid the tree. The persimmon is commonly propagated through cleft grafting and whip grafting.

HARVEST AND USE In Florida, the fruit ripens in the late summer and fall. It is clipped from the tree, leaving the calyx and a short length of stem intact. The fruit should be harvested when fully colored. It may be stored at room temperature for several days or several weeks, depending on the cultivar and other factors. Refrigeration will extend the life of astringent types.

The emerging flower of the persimmon is similar to that of the black sapote.

When these orange orbs appear in orchards around Gainesville, it is a sure sign of the approach of cooler temperatures.

Pineapple

SCIENTIFIC NAME: *Ananas comosus*
FAMILY: Bromeliaceae
OTHER COMMON NAME: Piña (Spanish)

Fruiting Calendar

JAN	FEB	MAR	APR	MAY	JUN	JUL	AUG	SEP	OCT	NOV	DEC

Characteristics

Overall Rating	★★★
Ease of Care	★★
Taste/Quality	★★★★★
Productivity	★★
Landscape Value	★★★★
Wind Tolerance	★★★★
Salt Tolerance	★★
Drought Tolerance	★★★
Flood Tolerance	★★
Cold Tolerance	★★

The pineapple is a major tropical fruit: it is versatile, richly flavored, and popular worldwide. The plant is relatively easy to grow and lends itself to many uses within the home landscape. However, because the pineapple has a long fruiting cycle and because it bears only one fruit at a time, the homeowner would need to establish many plants to enjoy this fruit on a regular basis.

Known Hazards
The spines along leaf margins and the stiff point at the leaf apex can cause mechanical injury.

182

GEOGRAPHIC DISTRIBUTION The pineapple is indigenous to Paraguay and Brazil. It spread to Central America and the Caribbean prior to the arrival of the Europeans. From 1860 until about 1920, the pineapple was an important commercial crop in south Florida. Production was reduced by a series of freezes and competition from Cuba.

PLANT DESCRIPTION The pineapple is a herbacious monocot. It grows to between 3 and 5 feet in height with an equal spread. A rosette of stiff, swordlike leaves surrounds the inconspicuous stem. The leaves often have spines along the margins. The root system is shallow, concentrated in the first 6 inches of soil. The plant makes an attractive ornamental and works well as a foundation planting. However, the sharp leaves present a hazard near walkways and play areas. Tiny purple-red flowers appear in a spike borne atop the terminal bud. After the initial fruit is harvested the plant may go on to produce 1 or 2 additional fruit on suckers or ratoons. Thereafter, the plant tends to decline.

FRUIT CHARACTERISTICS The fruit is compound, composed of the merged ovaries from many individual flowers. It is usually cone shaped or ovoid and may weigh from a few ounces to 20 pounds. The rind is divided into hexagonal plates called eyes. The flesh is white to yellow, juicy, sweet, and acidic. A fibrous core runs through the center of the flesh.

CULTIVARS Several cultivars are suitable for planting in Florida. 'Red Spanish' has proven satisfactory and produces a high quality fruit. 'Abacaxi' is disease resistant, produces a white-fleshed fruit of excellent flavor, and is favored for fresh consumption. 'Sugarloaf' related to 'Abacaxi,' bears a fruit of superb quality. Flesh is white or light yellow and very sweet. 'Smooth Cayenne,' the pineapple most often encountered in the supermarket, is marginally less well adapted to growth in Florida than 'Red Spanish.' 'Queen' bears a small, high-quality fruit on a compact plant.

RELATIVES The pineapple is the only important fruiting member of the Bromeliaceae or Bromeliad family, which contains over 2,000 species. Numerous species are native to Florida, including air plants such as Spanish moss, *Tillandsia usneoides.*

CLIMATE The pineapple will usually survive a temperature drop to 29° F. However, a temperature drop to 27° F will often prove fatal.

CULTIVATION The pineapple has moderate to high maintenance requirements. It prefers sandy, acidic soil. Good drainage is critical. The plant is drought tolerant, but benefits from periodic irrigation during the latter phases of fruit development. The pineapple has a high nitrogen requirement. Most growers apply a 8–3–8 or similar composition fertilizer at 3-month intervals. The plant also responds well to foliar sprays of urea. The pineapple can be forced to fruit through foliar sprays of ethylene-producing compounds.

Nematodes are a significant problem and can reduce the lifespan of the plant. Mealybugs and mites are more easily detected and controlled. Raccoons and other mammals sometimes gouge out the fruit immediately before it comes ripe. This is disheartening given the pineapple's long maturation period.

The pineapple is propagated through vegetative growth taken from the parent plant. The crown, removed from a harvested fruit, should be allowed to dry for several days in a shady location, after which it may be replanted. Slips, originating from the stalk below the fruit, suckers, originating from among the leaves near the base of the plant, and ratoons, originating from the soil near the base of the plant, may also be separated and replanted.

HARVEST AND USE The plant produces a single fruit during each cycle. Peak production occurs over the summer and early fall. The fruit typically undergoes a subtle color change at maturity. When tapped, a mature fruit will give off a flat, solid sound. The fruit can be refrigerated for up to 3 weeks but must be consumed quickly once it is returned to room temperature.

Few plants are more decorative than the pineapple. Along with the coconut palm and the banana, it serves as a symbol of the tropics.

Pitaya

SCIENTIFIC NAMES: *Hylocereus* spp. and *Selenicereus megalanthus*
FAMILY: Cactaceae
OTHER COMMON NAMES: Dragon Fruit, Night-Blooming Cereus

Fruiting Calendar

JAN	FEB	MAR	APR	MAY	JUN	JUL	AUG	SEP	OCT	NOV	DEC

Characteristics

Overall Rating	★★★
Ease of Care	★
Taste/Quality	★★
Productivity	★★★
Landscape Value	★★★
Wind Tolerance	★★
Salt Tolerance	★★
Drought Tolerance	★★★
Flood Tolerance	★★
Cold Tolerance	★★

The pitaya is a climbing cactus that bears a magnificent flower and colorful fruit. The fruit is rapidly increasing in popularity and commands a high price at the market. The flavor is good, although not everyone agrees that the fruit merits the degree of attention that it has garnered. The plant requires some maintenance. Whether it is grown for its fruit or its flowers, the pitaya makes an interesting addition to the home garden.

Known Hazards
The plant has sharp spines and has shown some invasive tendencies.

GEOGRAPHIC DISTRIBUTION The pitaya is native to Central America.

PLANT DESCRIPTION The pitaya is a vinelike, climbing cactus. Stems measure 2 to 4 inches in diameter, and are roughly triangular in cross section. Each areole harbors several sharp spines. The stems are divided into segments and may branch heavily. The plant will spread along the ground, but prefers to arch itself over a structure. It clings to almost any support by way of tough aerial roots. The pitaya has a quick growth rate and may begin bearing in as little as 2 years. Four-year-old plants are considered mature.

The showy flowers, which open at night, measure up to 9 inches in diameter. Many species and cultivars within the genus *Hylocereus* are self-compatible. However, a few require cross-pollination from a different cultivar. Hand pollination is widely practiced. This is accomplished by using a fine brush to transfer pollen from the anthers of one flower to the stigma of another.

FRUIT CHARACTERISTICS The fruit is oblong, measuring from 2 1/2 to 4 1/2 inches in length. The skin is smooth, punctuated by dull scales pointing toward the apex. Skin color may range from pink to red to yellow. Within the inedible rind is the moist pulp, which may be white, bright red, purplish-red, or pink. The flavor is mild and sweet. Numerous small black seeds, present in the flesh, do not interfere with eating.

SPECIES AND CULTIVARS Several species of climbing cacti from tropical America are referred to by the common name pitaya, including various members of the genera *Hylocereus* and *Stenocereus*. The yellow pitaya, *Selenicereus megalanthus,* bears knobby, yellow-skinned fruit. At least 50 named cultivars of pitaya exist, including 'Golden Dragon,' which has deep yellow skin and white interior flesh; 'Vietnam Dragon,' which produces a large fruit of high quality with deep-red skin and red flesh; 'Thai Dragon,' which has red skin and white flesh; 'Katom,' a high-quality cultivar tested in Israel; and 'Equador.' The cultivar 'Greeland' is self-fertile and produces adequate crops without hand pollination.

RELATIVES Many members of the family Cacaceae produce fruit. The Peruvian apple cactus, *Cereus repandus,* native to South America, is sometimes grown as an ornamental in Florida. It is a ribbed, columnar cactus that bears an aromatic fruit of fine quality. It can survive temperatures as low as 22° F. A similar species, *Cereus jamacaru,* is native to northeastern Brazil. The Barbados gooseberry, *Pereskia aculeata,* is a leafy cactus, which forms a climbing shrub. It is native to the Caribbean and northern South America. Leaflike sepals protrude from the skin of the immature fruit. The flesh is juicy and pleasantly tart. Members of the Cactaceae family native to Florida include the threatened triangle cactus, *Acanthocereus tetragonus;* the endangered pricklyapple cactus, *Harrisia aboriginum;* the Indian River pricklyapple, *Harrisia fragrans;* the endangered Key tree cactus, *Pilosocereus polygonus;* and several species from the genus *Opuntia*.

CLIMATE The pitaya prefers a dry, frost-free climate. Brief temperature drops to 29° F do not inflict significant harm. However, the plant will show stem blistering and other damage at 27° F. At lower temperatures, the stems liquefy and the plant may be killed to the ground.

CULTIVATION The pitaya prefers sandy, well-drained soil, and performs poorly on soils with high organic content. A trellis system or other support is essential. The goal is to encourage a spray of hanging or dangling branches, as it is on such branches that fruit production occurs. The pitaya is most often propagated through cuttings, obtained by severing foot-long lateral branches at a stem segment.

HARVEST AND USE In Florida, the pitaya typically fruits from May through September. An individual plant may bear more than 50 fruit per season. The fruit matures about 40 days after flowering and should not be harvested until it has achieved full coloration.

The glorious flower of the pitaya only opens at night.

The internal color of the fruit varies with the cultivar and species.

Pitomba

SCIENTIFIC NAME: *Eugenia luschnathiana*
FAMILY: Myrtaceae

Fruiting Calendar

JAN	FEB	MAR	APR	MAY	JUN	JUL	AUG	SEP	OCT	NOV	DEC

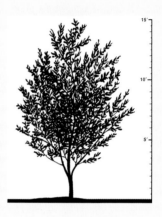

Characteristics

Overall Rating	★★★
Ease of Care	★★★★
Taste/Quality	★★★
Productivity	★★★
Landscape Value	★★★★
Wind Tolerance	★★★★
Salt Tolerance	★
Drought Tolerance	★★★
Flood Tolerance	★★★
Cold Tolerance	★★

The pitomba is an attractive, small tree that produces an abundance of yellow, aromatic, cherrylike fruit. However, it is rarely seen outside of collections. While the pitomba is not likely to achieve commercial status in Florida, it makes an excellent dooryard crop and is worth planting on a broader scale. The tree is easy to care for and is relatively free of problems. However, the fruit is subject to attack by the Caribbean fruit fly.

Known Hazards

None

GEOGRAPHIC DISTRIBUTION The species is native to Brazil. It has been grown in south Florida since about 1914.

TREE DESCRIPTION The pitomba is a small evergreen tree, capable of attaining a height of about 25 feet. The crown is densely foliated and of medium texture. Leaves are opposite, lanceolate, measuring between 2 and 3 inches in length. They are dark green, glossy, and decorative. The pitomba has a short, single trunk, covered with flaking gray bark. The pitomba makes a handsome landscape specimen. It is sometimes used as a loose screen or as an accent plant. The rate of growth is slow. White flowers, measuring about 3/4 of an inch in diameter, are born in great abundance from late April through May. They are typical of flowers produced by other trees within the *Eugenia* genus, with 4 petals and a prominent tuft of white stamens.

FRUIT CHARACTERISTICS The yellow fruit are obovate or slightly pyriform, measuring about 1 1/2 inches in diameter. The skin is smooth, thin, and glossy. Persistent sepals are present on the apex. The flesh is juicy and soft and encloses a single large seed or several smaller seeds. The flavor is pleasant and aromatic, with hints of cherry, balsam, and apricot.

CULTIVARS No cultivars have been selected in Florida.

RELATIVES Other members of the *Eugenia* genus discussed within this book include the cherry of the Rio Grande, *Eugenia aggregata;* the grumichama, *Eugenia brasiliensis;* and various stoppers, *Eugenia* spp. The Surinam cherry or pitanga, *Eugenia uniflora,* is an extremely common hedge plant in south Florida. It is native to South America. The fruit may be red or black. It has 7 or 8 ribs, giving it the appearance of a tiny pumpkin or Japanese lantern. At its best, the fruit is mild and slightly resinous. Because the Surinam cherry is classified as an invasive exotic, and because the fruit is mediocre, it is not recommended as a fruit tree. Other *Eugenias* that bear edible fruit include the arazá or araçá-boi, *Eugenia stipitata,* of the Amazon region; the cabellula, *Eugenia tomentosa,* of Brazil; the Cedar Bay cherry, *Eugenia reinwardtiana,* of Australia; the granadillo, *Eugenia ligustrina,* of Guiana and the West Indies; the guabiyu, *Eugenia puncens,* of Argentina; the murta, *Eugenia biflora,* of southern Mexico and Central America; the nanica, *Eugenia nhanica,* of Brazil; the ohia, *Eugenia malaccensis,* of Malasia; the pera de agua, *Eugenia magdalensis,* of Brazil; the pera do campo, *Eugenia klotzschiana,* of Brazil; the tatu, *Eugenia supraaxillaris,* of Brazil; the turtle berry, *Eugenia patrisii,* of northern South America;

and the uvalha, *Eugenia uvalha,* of Brazil. Other more distant relatives of the pitomba treated within this book include the blue grape, *Myrciaria vexator;* cattley guava, *Psidium cattleianum;* feijoa, *Feijoa sellowiana;* guava, *Psidium guajava;* and jaboticaba, *Myrciaria* spp.

CULTIVATION The pitomba is a low-maintenance tree. It will grow on most soils. It requires irrigation during establishment and during periods of drought. The tree, like its cousin the cherry of the Rio Grande, may suffer from an unexplained dieback. This is rarely fatal. Maggots of the Caribbean fruit fly sometimes ruin the fruit. The plant is grown primarily from seed in Florida. Seeds germinate in 2 to 5 weeks. Veneer grafting and cleft grafting have been carried out with some success.

HARVEST AND USE In Florida, the fruit typically ripens in July or August. The pitomba is usually consumed fresh. It can also be made into jellies or preserves.

The Surinam cherry—a relative of the pitomba—comes in a red-fruited form (top), which is harsh and resinous, and a black-fruited form (bottom), which is mild and sweet. The pitomba is a vastly superior fruit.

Prickly Pear

SCIENTIFIC NAME: *Opuntia ficus-indica*
FAMILY: Cactaceae
OTHER COMMON NAMES: Cactus Pear, Tuna (Spanish)

Fruiting Calendar

JAN	FEB	MAR	APR	MAY	JUN	JUL	AUG	SEP	OCT	NOV	DEC

Characteristics

Overall Rating	★★★
Ease of Care	★★★★★
Taste/Quality	★★★
Productivity	★★★
Landscape Value	★★★
Wind Tolerance	★★
Salt Tolerance	★★★★
Drought Tolerance	★★★★★
Flood Tolerance	★★
Cold Tolerance	★★★

The prickly pear bears a refreshing fruit with a sweet, melonlike flavor. Unfortunately, the fruit is difficult to harvest and prepare due to the presence of minute, barbed spines. The plant is easy to grow, produces showy flowers, and makes an attractive addition to the landscape. The hardiness and versatility of this cactus warrant its inclusion in the home garden.

Known Hazards

Tiny, barbed glochids embed themselves in the skin and cause irritation and itching. Sharp spines are also present on most varieties.

188

GEOGRAPHIC DISTRIBUTION Many species of prickly pear are native to the southern United States, Mexico, and portions of the Caribbean. The most important fruiting species is *Opuntia ficus-indica,* native to Mexico.

PLANT DESCRIPTION The prickly pear is characterized by cladodes—fleshy, flat oval pads that are actually flattened stems. Most varieties have spines. More problematic than the spines are the tiny barbs called glochids, which are hidden in the fuzzy aeroles dotting the surface of each pad. They readily detach from the plant and work their way into the skin like porcupine quills. The flowers are showy and variable. They may be yellow, red, violet, or purple. Flowering usually takes place in the spring.

The flower of the prickly pear varies in form and color but is always showy.

FRUIT CHARACTERISTICS The fruit is actually a fusion between stem tissue and the mature ovary. The fruit skin produces areoles and glochids. The juicy flesh may be bright pink, yellow, green, purple, orange, or white. While American consumers seem to prefer red or purple-fleshed varieties, Mexican consumers are partial to those with white or light green flesh. The flavor is mild and sweet, reminiscent of watermelon and strawberry.

CULTIVARS Most clones sold in Florida are unnamed. Important cultivars in Mexico include 'Reyna,' the leading commercial variety, which bears green fruit with light green flesh; 'Cristalina,' a midseason cultivar that bears large fruit of good quality; 'Naranjona,' a cultivar that bears yellow fruit with orange flesh; 'Chapeada,' a green-fruited cultivar; 'Amarilla Redonda,' a cultivar that bears orange fruit with yellow flesh; and 'Roja Pelona,' a cultivar that bears reddish-green fruits with deep red flesh. Researchers at Texas A&M University have selected a number of clones with desirable characteristics.

RELATIVES The *Opuntia* genus contains about 300 species, at least 25 of which produce edible fruit. The genus contains prickly pears, which have flat, beaver-tail-like pads, and chollas, which have cylindrical, jointed stems. Species native to Florida include the shell mound prickly pear, *Opuntia stricta,* a threatened species found along both coasts of the peninsula; the eastern prickly pear, *Opuntia humifusa,* a widely distributed species that is a favorite food of the endangered gopher tortoise; the bullsucker, *Opuntia cubensis;* the cockspur prickly pear, *Opuntia pusilla;* the Keys joe-jumper, *Opuntia triacanthos;* and the endangered semaphore cactus, *Opuntia corallicola.*

Two other cacti that produce worthwhile fruit are the strange Barbados gooseberry (top) and the stately Peruvian apple cactus (bottom).

CULTIVATION The prickly pear requires little cultural attention. It prefers well-drained, sandy or rocky soils. For maximum production, the plant must receive full sun. The cactus moth, native to South America, is a major pest. This insect was used as a biological control to kill off invasive *Opuntia* cacti in Australia and elsewhere. Its presence in Florida was first detected in 1989. The cactus moth lays its eggs in a spinelike "eggstick." Upon hatching, the larvae burrow into the soft tissue of the host plant. Monitoring the plants and removing the eggsticks will prevent the larvae from damaging the plant interior. Applying contact insecticides will kill the eggs and larvae.

The preferred method of propagation is by division. The top half of a mature pad is cut from the parent plant. This should be allowed to dry for a week until the cut calluses over. The pad should then be planted, with the base buried about an inch beneath the soil surface. Most will begin to root and form buds within 2 months.

HARVEST AND USE From the time of flowering, the fruit requires between 100 and 120 days to ripen. It should be picked when fully colored. Thick gloves are worn during harvest. As the first step in preparation, glochids should be removed from the fruit by rubbing it over burlap or other rough material. Average shelf life ranges from 10 days to 2 weeks. Young cactus pads are an important vegetable in Mexico, where they are referred to as nopales. The pads may be grilled, fried, used in chili, or eaten raw in salads.

Pomegranate

SCIENTIFIC NAME: *Punica granatum*
FAMILY: Punicaceae
OTHER COMMON NAME: Granada (Spanish)

Fruiting Calendar

JAN	FEB	MAR	APR	MAY	JUN	JUL	AUG	SEP	OCT	NOV	DEC

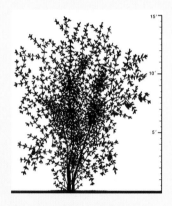

Characteristics

Overall Rating	★★★
Ease of Care	★★
Taste/Quality	★★★
Productivity	★★
Landscape Value	★★★★★
Wind Tolerance	★★★★
Salt Tolerance	★★★
Drought Tolerance	★★★
Flood Tolerance	★★
Cold Tolerance	★★★★

The pomegranate is a unique fruit, revered as a delicacy since ancient times. True, it is difficult to prepare and the juice leaves an indelible stain, but the flavor is exquisite. The tree has outstanding ornamental value with delicate foliage and showy, trumpet-shaped flowers. It can survive drought, freezing temperatures, and poor soil. However, it has some difficulty coping with the wet and humid conditions that occur during the summer in Florida.

Known Hazards

The plant is armed with sharp spines.

GEOGRAPHIC DISTRIBUTION The pomegranate is native to the Middle East. It has been cultivated since ancient times throughout the Mediterranean region. Spanish settlers may have brought the pomegranate to St. Augustine in the sixteenth century, although no evidence remains of such early introductions.

TREE DESCRIPTION The pomegranate is a small tree or large shrub. Under ideal conditions it may reach 25 feet in height. The narrow green leaves are lanceolot, measuring from 2 to 4 inches in length. Young foliage is reddish, pink, or bronze. Leaf margins are wavy. Small branches are armed with spines. The pomegranate makes an outstanding landscape specimen. The tree bears showy trumpet-shaped flowers that may exceed 3 inches in length. These may be red, orange, or white.

FRUIT CHARACTERISTICS The fruit is a globose, atypical berry. It measures between 3 and 5 inches in diameter. The rind is tough and leathery. The skin, which ranges in color from green to yellow, is overlaid by a red blush, which intensifies as the fruit matures. The interior is divided into chambers by white membranes that are bitter and inedible. Each chamber is densely packed with seeds surrounded by an aril. The aril is swollen with translucent, red pulp. The edible portion comprises less than 40 percent of the overall weight of fruit. The flavor is lightly astringent, slightly sweet, refreshing, and addictive. An individual fruit may contain anywhere from 500 to 900 seeds. The seeds are small and angular and range in texture from hard to soft and chewable.

CULTIVARS The ability of a cultivar to fruit under wet and humid conditions is crucial. 'Vietnam' is well suited to south Florida and central Florida. The tree bears a large crop from March through May, before the advent of the rainy season. The fruit is small and of fair quality. 'Plantation Sweet,' selected in Georgia from a tree said to be more than 100 years old, is prolific, cold hardy, and bears a medium-large fruit that ripens in November. 'Big Red,' a soft-seeded variety, has proven somewhat productive in Florida. 'Francis,' originally from Jamaica, may have potential in south Florida in light of its origin in the humid tropics. Other promising cultivars include 'Fleishman,' 'Purpleseed,' 'Spanish Ruby,' and 'Christine.' 'Wonderful' an important commercial variety in Califonia, was originally selected in Florida, but is only marginally adapted to Florida growing conditions.

RELATIVES The pomegranate belongs to the tiny family Punicaceae. The genus *Punica* is coextensive with the family. The only other member of the family and genus is *Punica proto-punica,* which is only found on Socotra Island, located in the Indian Ocean just east of the horn of Africa.

CLIMATE The pomegranate is a tree of subtropical and warm-temperate climates. Depending on the cultivar, it may be injured at temperatures ranging from 20° to 5° F. It prefers dry summer conditions. In south Florida, the pomegranate is most productive following a cold winter.

CULTIVATION The pomegranate is a low-maintenance tree. It has a high degree of drought tolerance. It is not particular as to soil type, but requires good drainage. The tree should be trained to form a vaselike shape atop a short trunk. Suckers forming on the lower trunk should be removed. Few serious pests bother the pomegranate in Florida. However, anthracnose and other fungal diseases can cause extensive problems during wet weather. The pomegranate is often reproduced through cuttings. Grafting is ineffective because the plant tends to sucker below the graft union.

HARVEST AND USE The fruit matures from 5 to 7 months after flowering, usually from mid-summer through fall. Color is the critical factor for determining ripeness. The pomegranate is often consumed out of hand. The fruit is often quartered and the pulp sacs removed. The juice can be strained or pressed from the fruit. Grenadine is a reduced juice made from pomegranate and sugar. Most commercial products no longer use real pomegranate juice. The pomegranate has an exceptionally long shelf life, especially when refrigerated.

Shown here midway between flower and fruit, the pomegranate is a treat for the eyes at every phase of development.

An immature fruit of the cultivar 'Vietnam.' The arils surrounding the seeds are just beginning to swell with juice.

Sapodilla

SCIENTIFIC NAME: *Manilkara zapota*
FAMILY: Sapotaceae
OTHER COMMON NAMES: Chicku, Naseberry, Chicle,
Nispero or Zapotilla (Spanish)

Fruiting Calendar

JAN	FEB	MAR	APR	MAY	JUN	JUL	AUG	SEP	OCT	NOV	DEC

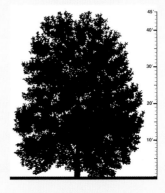

Characteristics

Overall Rating	★★★★★
Ease of Care	★★★★
Taste/Quality	★★★★
Productivity	★★★★
Landscape Value	★★★★
Wind Tolerance	★★★★★
Salt Tolerance	★★★★
Drought Tolerance	★★★★★
Flood Tolerance	★★★★
Cold Tolerance	★★

The sapodilla is a handsome tree, capable of enduring wind, drought, and salinity. It fruits abundantly over much of the year. The fruit is delicious and is an important commercial crop in Latin America and elsewhere. The species suffers two shortcomings. First, it is intolerant of winter cold. Second, it is considered an invasive exotic in Florida. However, the excellence of the fruit and many other attributes make it difficult to resist the allure of the sapodilla.

Known Hazards

The tree is an invasive exotic. The seeds present a choking hazard. In addition, the seed kernel is reportedly toxic.

GEOGRAPHIC DISTRIBUTION The sapodilla is native to Central America and southern Mexico, including the Yucatan Peninsula. Wild populations still exist. The species has been exported to almost every tropical region and has become a popular commercial crop in many areas. While it remains an important fruit crop throughout its native range, India is currently the world's largest producer.

TREE DESCRIPTION The sapodilla is a medium-large evergreen. It will generally attain a height of 40 to 45 feet in Florida. It can grow much larger in its native habitat. The crown is dense, rounded, and somewhat symmetrical. The alternate leaves are ornate, glossy, leathery, and deep green. They are elliptic to lanceolate in form. Leaves are thickly clustered near the branch tips. They average 4 or 5 inches in length. In some cultivars, emerging leaves have a distinct pink or light red cast. The foliage presents a rich, textured appearance. The bark of mature specimens is furrowed. The wood is hard and durable. When the tree is cut, it exudes sticky white latex known as chicle. Once the prime ingredient in chewing gum, this substance has been replaced by synthetics. The root system is shallow and is concentrated within the first 2 feet of soil.

Emerging leaves of the 'Alano' sapodilla.

In the Redlands of Miami-Dade County, the sapodilla is planted as a windbreak to protect avocados, mangos, and lychees.

The tree is slow-growing but long-lived. Century-old specimens show no signs of decline. The sapodilla makes an extraordinary landscape specimen. It is an imposing shade tree. It also can be employed with good effect in a group planting or may be set out in a row as a windbreak or screen. The sapodilla produces inconspicuous, bell-shaped, bisexual flowers, measuring less than 1/2 inch in diameter. The outer sepals are light brown. The inner sepals are an off-white. There are 6 stamens. The flowers are carried on short stalks, which dangle from the leaf axils in loose groups.

The sapodilla has been classified as an invasive exotic. There is no question that the species has sprung up in unintended locations. However, there is also some evidence that the sapodilla is less invasive than its detractors would suggest. It is extremely uncommon in hammocks and undisturbed locations removed from suburban and agricultural areas. Trees from an ancient, long-abandoned sapodilla grove on No Name Key have not overwhelmed native foliage on the island. Seedlings have sprouted under the canopy, but few volunteers have sprung up beyond the boundaries of the original planting. Planting the species in a suburban setting would probably not contribute to a decline in native vegetation.

FRUIT CHARACTERISTICS The fruit is a large berry covered with brown scurf. It measures from 2 to 5 inches in diameter. It may be spindle shaped, globose, or obovate. The moist flesh ranges in color from brownish-yellow to reddish-brown. The central core is soft and edible. The fruit may contain up to 12 seeds, although better cultivars contain an average of between 2 and 5 seeds. The seeds are hard, shiny, dark brown to nearly black. They are tear-shaped, with a tiny hook at one corner and average about 3/4 of an inch in length. When perfectly ripe, the sapodilla ranks in the top echelon of all tropical fruit. The flavor is reminiscent of a luscious pear, suffused with brown sugar and scented with jasmine.

CULTIVARS More than 100 cultivars have been selected. Great disparity exists between the best varieties and inferior varieties. The following cultivars are recommended for Florida.

'Alano' - 'Alano' is a Hawaiian selection that produces small, spindle-shaped fruit of excellent quality. The flesh, colored a light golden brown, is finely textured and contains no grit. Production is high. The tree is beautiful with narrow leaves and a pyramidal growth habit. New flushes of growth are pink. The tree may fruit sporadically throughout the year, although its most productive period runs from December through July.

The 'Alano' sapodilla produces fruit of excellent quality and is well suited to dooryard planting.

'Hasya' - This cultivar bears fruit of excellent quality on a large, upright tree. The fruit is medium size. The flesh is reddish-brown. Hasya is the number-one commercial variety in several Mexican states. In Florida, the main season of production runs from January through April.

'Morena' - This cultivar closely resembles 'Hasya.' The fruit is of excellent quality. The flesh is smooth textured with brownish-red coloration. New shoots are bright green. In Florida, 'Morena' fruits most heavily from February to April.

'Tikal' - 'Tikal' is a seedling selection made in Florida. The fruit is elliptic in shape. Tikal is very productive and bears fruit of very fine quality.

'Brown Sugar' - This selection originated in Homestead in 1948. It bears medium-large fruit of approximately the shape and size of a beefsteak tomato. The fruit is very sweet, but granular in texture. The flesh is uniformly deep brown in color. 'Brown Sugar' is a good producer of high-quality fruit. If left on the tree too long, the fruit is susceptible to attack by the Caribbean fruit fly.

The cultivar 'Brown Sugar' is an old standard and produces a fruit of very good quality.

'Makok' - For those with limited space, 'Makok' may represent the best choice. This dwarf tree, which originated in the Philippines, produces small fruit of very good quality.

Other worthwhile cultivars include 'Gonzalez,' a Philippine selection that produces small to medium-size fruit of very good quality; 'Modello,' a moderately productive cultivar from Mexico; 'Molix,' a Mexican selection that bears medium-size fruit of very good quality; 'Oxkutzcab' a Mexican selection that produces large fruit of good flavor but with grainy texture; and 'Prolific,' a very productive commercial cultivar that produces fruit of fair to good quality. 'Russell,' a selection from the Florida Keys, lags in production, but bears a high-quality, pink-fleshed fruit.

RELATIVES The Sapotaceae family is composed of about 50 to 70 genera and about 1,000 species. Most are found in tropical, subtropical, and warm-temperate regions. Species described within this book, include the abiu, *Pouteria caimito;* canistel, *Pouteria campechiana;* mamey sapote, *Pouteria sapota;* and star apple, *Chrysophyllum cainito.*

The genus *Manilkara* contains several species that produce edible fruit, including the balata, *Manilkara bidentata,* of Central America and northern South America, and the maçaranduba, *Manilkara huberi,* of the Amazon region. More distantly related is the miracle fruit, *Synsepalum dulcificum,* a very minor fruit from West Africa that is occasionally grown as a curiosity in south Florida. The tiny red fruit trick the taste buds into perceiving sour flavors as sweet. For up to an hour after consuming a miracle fruit, the most acidic lime or lemon can be eaten with impunity.

Several fruiting relatives of the sapodilla are native to Florida. The closely related wild dilly, *Manilkara bahamensis,* native to south Florida and the Caribbean, produces a small fruit that is rarely consumed, except in survival situations. The safron plum, *Sideroxylon celastrinum,* native to central and southern parts of the peninsula, produces a sweet, edible fruit of up to 1 inch in diameter. The mastic, *Sideroxylon foetidissimum,* native to south Florida and the Caribbean, is a large tree that bears a small fruit of fair quality. The gum bumelia or woolly buckthorn, *Bumelia lanuginosa,* grows in the panhandle and northwestern parts of the peninsula. It produces 1/4- to 1/2-inch black fruit with sweet pulp and a single seed. However, the fruit is said to cause digestive upset if consumed in quantity. The tough bumelia, *Bumelia tenax,* native to the east coast of Florida, produces edible 1/2-inch black berries. The satinleaf, *Chrysophyllum oliviforme,* a threatened species native to south Florida, bears a purple berry measuring up to 3/4 of an inch in length, with sweet white flesh surrounded by rubbery skin.

CLIMATE The sapodilla is a tropical tree. It can withstand a subtropical climate so long as freeze events are infrequent and light. A mature tree may endure a temperature drop to about 26° or 27° F. The 1989 freeze killed or severely damaged several mature trees in the vicinity of Stuart, where the temperature dropped to 26° F for two successive nights.

CULTIVATION The sapodilla is a tough and undemanding tree. It is tolerant of most soil types, including the thin limestone soils of extreme southeastern Florida and the Keys. It rarely develops nutritional deficiencies. It can tolerate serious and prolonged drought. At the same time, the tree has shown some ability to withstand brief flooding. The sapodilla is able to withstand moderate salinity in the soil, water, and air. In the Keys, it can be found growing as little as 3 feet above mean high tide. The tree is remarkably wind resistant. In 1992, Hurricane Andrew swept across the Redlands Agricultural Area, flattening and splintering many acres of mangos, lychees, and avocados. However, many sapodillas survived Andrew's wrath with only superficial damage. The sapodilla's adaptability makes it well suited to many locations in which few other fruit trees will grow.

PESTS AND DISEASES The sapodilla suffers fewer pest and disease problems than most fruit trees. Aphids and white flies can cause minor damage. Some cultivars, if left on the tree too long, are moderately susceptible to attack by maggots of the Caribbean fruit fly. Root weevils may cause significant damage to a young tree. Squirrels, raccoons, and opossums sometimes raid the tree for its fruit.

PROPAGATION Cloned cultivars are greatly preferred to seedlings, as seedlings are extremely variable. Veneer grafting and wedge grafting have shown reasonable rates of success. Grafting is most successful during the summer, when the tree is in an active state of growth. Air layering has been practiced on occasion. Attempts to root leafy cuttings and hardwood cuttings have not produced satisfactory results.

HARVEST AND USE The sapodilla is prolific in its bearing habits. A 10-year-old tree may produce as much as 100 pounds per year. An older tree may produce in excess of 500 pounds of fruit. The fruit matures about 10 months after flowering.

The fruit exhibits few outward signs of maturity. Fruit picked prematurely do not ripen in a satisfactory manner. The most common method of determining maturity is to scratch away some of the exterior scurf to reveal the color of the skin below. If the skin is green, the sapodilla is not mature. If it is yellow, yellow-orange, or tan, the fruit is ready to pick. A fruit that separates easily from the stem, without leaking significant latex, is considered mature.

Mature fruit ripen in less than a week at room temperature. Peak ripeness only lasts about 2 days. If eaten too early, the fruit is gummy and sharply astringent. If eaten too late, the fruit is mawkish and mushy. When the fruit emits a slight, perfumed odor, and is slightly soft to the touch, it is at a perfect stage. The fruit is usually eaten out of hand, cut in half, with the flesh spooned from the skin and the seeds discarded. The seeds have a hooklike projection at one corner and present a choking hazard.

The bell-shaped flowers of the sapodilla are similar to those of its cousins, the abiu and the canistel.

Seagrape

SCIENTIFIC NAME: *Coccoloba uvifera*
FAMILY: Polyconaceae
OTHER COMMON NAME: Uva de Playa (Spanish)

Fruiting Calendar

JAN	FEB	MAR	APR	MAY	JUN	JUL	AUG	SEP	OCT	NOV	DEC

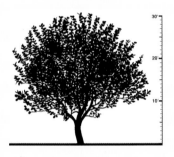

Characteristics

Overall Rating	★★
Ease of Care	★★★★
Taste/Quality	★★
Productivity	★★★
Landscape Value	★★★★
Wind Tolerance	★★
Salt Tolerance	★★★★★
Drought Tolerance	★★★
Flood Tolerance	★★★★
Cold Tolerance	★★

The seagrape is one of south Florida's most attractive native plants. It is tolerant of adverse conditions. It will grow in exposed, beachfront locations, where few other trees—much less fruit trees—can survive. The fact that it produces a useful fruit is an added bonus. The fruit does not rank among the elite fruits of the subtropics and is hardly palatable when fresh. However, it can be used to make an excellent jelly.

Known Hazards

None

196

GEOGRAPHIC DISTRIBUTION The seagrape is native to the Caribbean, the Bahamas, and south Florida. It grows along the northern coast of South America and along both coasts of Central America and Mexico.

TREE DESCRIPTION The seagrape is a small to medium evergreen tree and can attain a height of about 35 feet. It may also form a sprawling shrub or a multi-trunked thicket. The rounded leaves measure between 5 and 9 inches in diameter. The central vein is often crimson in color. The bark is somewhat guavalike in appearance, smooth in texture, with orange, tan, and gray patches. The wood is dense but somewhat brittle. Flowers appear on spikes or racemes, measuring between 8 and 12 inches in length. The flowers are small, off-white, and pleasantly fragrant. The sea grape is dioecious: male and female flowers are borne on separate trees. Although a male tree is required for pollination, only the female tree produces fruit. Flowers may appear sporadically throughout the year, but are most common from February through July.

FRUIT CHARACTERISTICS The fruit is round or oval, about 3/4 of an inch in diameter, and roughly grapelike in appearance. The apex is lightly grooved. The fruit is produced in elongated clusters that hang pendulously from the branches of the female tree. A cluster may contain as many as 70 fruit. The skin is light green at first, turning pink or purple when ripe. The flavor varies from tart and acidic to mild and sweet. Embraced within the pulp is a large, hard, elliptic seed, pointed on top. Fruit typically ripen during the summer or fall.

CULTIVARS Although the seagrape has been cloned for landscape purposes, no selection has occurred based on the plant's fruiting characteristics.

RELATIVES The seagrape is a member of the diverse Buckwheat or Polygonaceae family, which consists of a number of herbs and a few trees. The *Cocoloba* genus is composed of about 50 species, of which at least 15 bear edible fruit. The pigeon plum, *Coccoloba diversifolia,* is native to coastal hammocks in south Florida. The species forms a small tree of picturesque habit, which bears edible 1/2-inch drupes. The seaplum, *Coccoloba diversifolia* x *Coccoloba uvifera,* a hybrid between the seagrape and pigeon plum, is sometimes used for landscape purposes in Florida. Other seagrape relatives are scattered throughout the Caribbean.

CLIMATE The seagrape requires a tropical or subtropical climate. Mature specimens have survived temperature drops to 23° F. The tree grows as far north as Cape Canaveral on the east coast and Clearwater on the Gulf coast.

CULTIVATION The seagrape is a hardy tree and flourishes with little or no care. It is highly tolerant of salt in the water and soil. This characteristic makes it uniquely suited for growth on barrier islands. Indeed, the seagrape will grow in locations where the root system is occasionally immersed in salt water. It establishes readily in windy and exposed sites. The tree tends to break up in hurricane-force winds. While it makes best growth on sandy loam, the seagrape adapts readily to many soil types. It is drought tolerant. No serious pests or diseases affect the seagrape in Florida. Birds sometimes eat the fruit. The seagrape is usually grown from seed. Seeds sown at a shallow depth will germinate readily, usually between 3 and 6 weeks after they are planted. The plant can also be reproduced by air layering, by veneer grafting, and by rooting hardwood cuttings.

HARVEST AND USE A mature specimen will produce several gallons of fruit over the course of a season. The fruit should be harvested when fully colored. The fruit tend to drop shortly after ripening. The seagrape can be eaten out of hand when fully ripe, although the fresh fruit is of poor quality. More often it is used to make jellies or preserves.

Flowering peaks in May and June.

Star Apple

SCIENTIFIC NAME: *Chrysophyllum cainito*
FAMILY: Sapotaceae
OTHER COMMON NAME: Caimito (Spanish)

Fruiting Calendar

JAN	FEB	MAR	APR	MAY	JUN	JUL	AUG	SEP	OCT	NOV	DEC

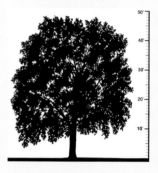

Characteristics

Overall Rating	★★
Ease of Care	★★★
Taste/Quality	★★★
Productivity	★★★
Landscape Value	★★★
Wind Tolerance	★★★★
Salt Tolerance	★
Drought Tolerance	★★★
Flood Tolerance	★★★
Cold Tolerance	★

The star apple is a common fruit throughout the Caribbean and much of Latin America. Those who casually sample the fruit often take to it immediately. The foliage is ornamental and the tree requires little care. However, the tree's sensitivity to cold is a limiting factor. In addition, the tree is far too large for the average suburban yard. In light of these factors, the star apple is probably a species best suited to the serious collector.

Known Hazards

None

198

GEOGRAPHIC DISTRIBUTION The star apple is thought to be indigenous to Central America. It has become naturalized on several islands of the Caribbean. The star apple is grown in many tropical and subtropical regions, most often as a dooryard tree.

TREE DESCRIPTION The star apple is a large tree, growing to 50 feet in south Florida. While usually classified as an evergreen, it may drop most or all of its leaves over the winter, especially if it is subjected to cold temperatures. Leaves are alternate, elliptic, and measure 3 to 6 inches in length. They are dark green above. The underside is coated with velvety bronze fuzz. Leaf margins are slightly rolled, and the leaves are somewhat stiff and leathery in texture. Flowers are small, purplish-white or yellow. They are grouped in the leaf axils. The tree is moderately fast growing and is capable of prodigious yields.

FRUIT CHARACTERISTICS The fruit is globose, measuring from 2 to 3 inches in diameter. The fruit is initially lime green, but may turn dark purple at maturity. The rind is thick, rubbery, and inedible. The pulp, forming several faint lobes, fills a cavity within the rind. The pulp is mucilaginous, white or slightly translucent. When halved, the seeds and pulp form a starlike pattern that radiates from the central core. The seeds are brown, measure about 3/4 of an inch in length, and are slightly flattened. They range in number from 3 to 10. The seeds are surrounded by a rubbery layer, which adheres to the seed coat but not to surrounding pulp.

CULTIVARS As noted, the star apple comes in green and purple variations. One well-regarded cultivar selected by rare-fruit pioneer William Whitman in Port-au-Prince, Haiti, and dubbed the 'Haitian Star Apple,' was introduced to Florida in 1952. A variety named 'Blanco Star,' with reddish-purple skin, is also available in south Florida. 'Philippine Gold' is a green-skinned cultivar.

RELATIVES The star apple is a member of the important Sapotaceae family. Related species discussed in this book include the abiu, *Pouteria caimito;* canistel, *Pouteria campechiana;* mamey sapote, *Pouteria sapota;* and sapodilla, *Manilkara zapota.* The satinleaf, *Chrysophyllum oliviforme,* is a native species related to the star apple. It produces a dark purple berry measuring up to 1 inch in length, with white flesh surrounded by rubbery skin.

CLIMATE The star apple is a tropical tree. It will suffer defoliation if exposed to freezing temperatures. A young tree may be killed by exposure to a brief overnight frost. An older tree will suffer limb injury or may die if the temperature drops to 28° F or below. The star apple is susceptible to cold injury in any part of the peninsula.

CULTIVATION Apart from its sensitivity to cold, the star apple is an undemanding tree. It requires well-drained soil. However, it will grow on infertile soil, sand, and the oolitic limestone soils of extreme south Florida. It will tolerate alkaline soils with a pH of up to 7.5. The star apple is drought tolerant. However, a young tree should be irrigated for several months until it is established. Irrigation during flowering may increase yields. A single application of a balanced fertilizer may benefit a tree growing on poor soil. The tree is sometimes grown from seed. It has been successfully propagated through air-layers, hardwood cuttings, and grafting.

HARVEST AND USE Harvest is labor intensive, as the fruit must be clipped from the upper branches of the tree—which may be 50 feet above the ground. A cherry picker may be required. The fruit mature over the late winter and spring. A star apple is considered ripe when rubbery and slightly soft to the touch. Fruit quality tends to be at its best following a warm, frost-free winter. The fruit is best served chilled. Most often, it is simply halved and the flesh scooped from the rind.

The diminutive flowers of the star apple form in the leaf axils.

Fruit of the 'Haitian Star Apple' cultivar is attractive and delicious.

Stopper

SCIENTIFIC NAMES: *Eugenia* spp. and *Myrcianthes fragrans*

FAMILY: Myrtaceae

Fruiting Calendar

JAN	FEB	MAR	APR	MAY	JUN	JUL	AUG	SEP	OCT	NOV	DEC

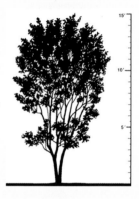

Characteristics

Overall Rating	★★★
Ease of Care	★★★★
Taste/Quality	★★
Productivity	★★★
Landscape Value	★★★★★
Wind Tolerance	★★★★
Salt Tolerance	★★
Drought Tolerance	★★★★
Flood Tolerance	★★★
Cold Tolerance	★★

The stoppers are valuable native plants, several of which produce edible berries. They are reasonably hardy in south Florida and present an excellent choice for home landscaping purposes. The fruit provide an added bonus. Among the 5 species commonly referred to as stoppers, Simpson's stopper is usually regarded as the most noteworthy fruit-bearing species.

Known Hazards

None

GEOGRAPHIC DISTRIBUTION The stoppers are native to south Florida and the Caribbean. Several species are rare and cannot be collected from the wild. However, most are available from nurseries specializing in native plants.

TREE DESCRIPTION Stoppers grow as medium to large shrubs with multiple trunks and a vase shape. However, they may also form small trees or may be trained as a hedge. The shrubs range in height from 10 to 25 feet. The canopy forms a fairly dense, rounded head. Leaves are simple, entire, and elliptic, typically measuring about 2 inches in length. The foliage is evergreen, of fine to medium texture. The bark is smooth and mottled, sometimes peeling, and ranges in color from reddish-brown to gray. Flowers are small, white to cream in color, and similar to those produced by other members of the *Eugenia* genus. Flowering usually peaks during the late spring and summer. Fruit and flowers are often present at the same time.

FRUIT CHARACTERISTICS The fruit is an aromatic berry, ranging in size from 1/4 to 1/2 inch. It is typically red or black in color and may be either oval or slightly pyriform in shape. The flavor of Simpson's stopper is sweet with hints of cedar and citrus. However, the fruit quality is somewhat variable.

SPECIES Among the stoppers, Simpson's stopper, *Myrcianthes fragrans,* produces the best fruit. It is an attractive landscape plant, forming a medium to large shrub with peeling rust-colored bark. It is slightly more cold tolerant than several of its cousins. Simpson's stopper produces bright orange-red fruit that ranges in quality from fair to very good. It is considered a threatened species in Florida. The Spanish stopper, *Eugenia foetida,* typically attains a height of 12 to 15 feet. The fruit, although small, are numerous and edible. They are dark red to black, usually maturing in early fall. The white stopper, *Eugenia axillaris,* is a medium-size shrub with light gray bark. The foliage has a musky aroma and new foliage has a reddish tint. Flowers are white to yellowish-white to pinkish-white. The small, purple berries are edible but of little value. The red berry stopper, *Eugenia confusa,* and the red stopper, *Eugenia rhombea,* are considered endangered plants in Florida.

RELATIVES The *Eugenia* genus, made up of about 400 species, is part of the Myrtaceae family. Several species from this genus are planted for their fruit. These include the cherry of the Rio Grande, *Eugenia aggregata;* the pitomba, *Eugenia luschnanthiana;* the Surinam cherry, *Eugenia uniflora;* and the grumichama, *Eugenia braziliensis.* Within the genus *Myrcianthes,* to which the Simpson's stopper belongs, is the poorly known guabiyu, *Myrcianthes pungens,* native to mountainous regions of South America. This large tree produces a fruit said to resemble that of the strawberry guava. More distant relatives include the feijoa, *Feijoa sellowiana;* the guava, *Psidium guajava;* and the jaboticaba, *Myrciaria* spp.

CLIMATE Stoppers are tropical to subtropical in their climatic requirements. Simpson's stopper is capable of surviving in protected locations as far north as St. Augustine. The Spanish stopper and white stopper will grow in coastal areas as far north as Cape Canaveral on the east coast and the Tampa Bay region on the west coast. The red stopper and the redberry stopper are limited to coastal areas of extreme south Florida.

CULTIVATION Once established, the stopper is a very low-maintenance plant. It is drought tolerant and moderately salt tolerant. It grows well on many soil types and can adapt to a wide pH range. The stopper has few serious pests or diseases. The lobate lac scale, an insect pest recently introduced into Florida, has been found to infest the foliage. The Caribbean fruit fly occasionally infests ripe fruit. The plant is propagated almost exclusively from seed.

HARVEST AND USE The fruit ripens over the summer. It is collected as soon as it attains full coloration. It can be eaten fresh or used in preserves.

Simpson's stopper has beautiful foliage and decorative flowers and is capable of bearing fruit of good quality.

Strawberry

SCIENTIFIC NAME: *Fragaria ananassa*
FAMILY: Rosaceae
OTHER COMMON NAME: Fresa (Spanish)

Fruiting Calendar

JAN	FEB	MAR	APR	MAY	JUN	JUL	AUG	SEP	OCT	NOV	DEC

Characteristics

Overall Rating	★★★★
Ease of Care	★★★
Taste/Quality	★★★★
Productivity	★★★
Landscape Value	★★★
Wind Tolerance	★★★
Salt Tolerance	★★
Drought Tolerance	★★★
Flood Tolerance	★★
Cold Tolerance	★★★

The strawberry is a popular spring crop and represents a valuable addition to any Florida garden. The plant requires little maintenance apart from periodic irrigation. It consumes little space within the garden. It also has the advantage of bearing fruit within a few months after planting. The strawberry will grow in all regions of the state. However, it is regarded as an annual crop and is replanted at the start of each growing season.

Known Hazards

Some are severely allergic to the fruit.

GEOGRAPHIC DISTRIBUTION Several species of strawberry have evolved simultaneously in distant geographic regions. The cultivated strawberry, which originated in France, is the result of an accidental cross between *Fragaria virginiana,* from eastern North America, and *Fragaria chiloensis,* from the Pacific coast of South America. Despite its poor handling characteristics and perishable nature, the strawberry is an important commercial crop. The United States is the world's largest producer. In Florida, production is concentrated in Hillsborough and Manatee Counties.

PLANT DESCRIPTION The strawberry is a creeping, perennial herb, rarely attaining a height of more than 18 inches. It is usually grown as an annual in Florida. The dark green leaves are trifoliate. The plant spreads by way of low runners. Strawberry flowers are bisexual and are generally self-pollinated. The flowers are white, about 1 inch in diameter, composed of 5 petals and about 25 stamens coated with yellow pollen.

FRUIT CHARACTERISTICS Technically, the 'fruit' of strawberry is neither a berry nor a fruit. It is an aggregate of achenes. The tiny seeds on the outside of the skin comprise the actual "fruit." A ripe strawberry typically measures between 1 and 2 1/2 inches in length. The fruit ripens between 25 and 40 days after flowering.

CULTIVARS While more than 100 strawberry cultivars have been bred and selected, only a scattering are well adapted to Florida's climatic conditions. Cultivars recommended for home garden planting in Florida include 'Sweet Charlie,' 'Florida Belle,' 'Florida 90,' 'Camarosa,' and 'Chandler.' Other cultivars that may be suitable for planting in Florida include 'Carlsbad,' 'Tioga,' 'Earlibright,' 'Oso Grande,' 'Rosa Linda,' 'Strawberry Festival,' 'Sequoia,' 'Selva,' and 'Treasure.'

RELATIVES The strawberry, like the blackberry, belongs to the subfamily Rosoideae within the Rosaceae family. The *Fragaria* genus consists of about 35 species, distributed throughout the Americas, Europe, and Asia. Other members of the Rosaceae family presented within this book include apple, *Malus domestica;* blackberry, *Rubus* spp.; capulin, *Prunus salicifolia;* chickasaw plum, *Prunus angustifolia;* loquat, *Eriobotrya japonica;* mayhaw, *Crataegus* spp.; peach, *Prunus persica;* and pear, *Pyrus* spp.

CLIMATE The strawberry is at home in all areas of Florida, although frost may damage tender flowers and fruit in some locations. If flowers or fruit are lost as a result of cold weather, the plant will usually produce a second flush of flowers.

CULTIVATION The strawberry has moderate maintenance requirements. The plant is usually set out in October or November, with 10- to 18-inch spacing. It is often grown in mounded rows. It produces flowers over the winter months, bears fruit in the spring, and sends out runners over the summer. The strawberry prefers well-drained, sandy soil. Black plastic mulch is frequently used both in commercial operations and in home gardens to control weeds, minimize evaporation, and warm the soil. The plant has a shallow root system and requires irrigation during dry periods. Birds frequently steal the fruit. Some growers paint stones red and scatter these throughout the patch early in the season, theorizing that birds will soon come to regard all red objects as stones. Thrips and mites are often present, but are ignored until populations become troublesome. The strawberry can be propagated by division—by digging up and replanting new plants formed at the ends of runners. However, most growers begin with fresh, disease-free plants purchased at the start of each growing cycle.

HARVEST AND USE In Florida, strawberry production begins in January, peaks in March, and tapers off by May. Each plant is capable of producing 1 or 2 pints of fruit over the course of a season. Fruit is considered ready to pick when 3/4 or more of the skin surface is red. The fruit does not sweeten once picked and stores for only a few days unless frozen or otherwise preserved.

At all phases of development, the seeds of the strawberry are located on the outside of what is commonly thought of as the fruit.

Sugar Apple

SCIENTIFIC NAME: *Annona squamosa*
FAMILY: Annonaceae
OTHER COMMON NAMES: Sweetsop, Anon de Azúcar
(Spanish)

Fruiting Calendar

JAN	FEB	MAR	APR	MAY	JUN	JUL	AUG	SEP	OCT	NOV	DEC

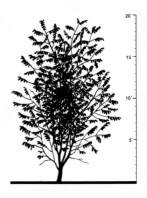

Characteristics

Overall Rating	★★★
Ease of Care	★★★
Taste/Quality	★★★
Productivity	★★★
Landscape Value	★★★
Wind Tolerance	★★★
Salt Tolerance	★★
Drought Tolerance	★★★
Flood Tolerance	★★★
Cold Tolerance	★

The sugar apple is a well-regarded member of the Annona family. It is popular as a dooryard crop throughout the American tropics. The pulp is sweet and delicious. Because of its reliable production, overall quality, and the small size of the tree, the sugar apple makes an outstanding backyard specimen for those living in coastal south Florida. However, the tree is cold tender and the fruit is subject to attack by the chalcid wasp.

Known Hazards

The seed contains toxins. The sap may be an irritant.

GEOGRAPHIC DISTRIBUTION The sugar apple is native to tropical America, although its exact point of origin is not known. It is widely cultivated throughout the tropical lowlands of South America, Central America, Mexico, and the Caribbean.

TREE DESCRIPTION The sugar apple is a small, briefly deciduous tree reaching about 15 or 20 feet in height. The branches are sparse and the canopy is open, rounded but irregular. The sugar apple has smooth, gray-brown bark. Leaves are alternate, fleshy, lanceolate to oblong, averaging 3 to 5 inches in length. They are dull green above, pale green below. The landscape quality of the sugar apple is good. The growth rate is moderately fast. Seedlings usually begin to bear within 3 or 4 years. The tree has a productive life of about 15 years. Flowers appear in the spring and early summer, usually on new growth. The flower, about 1 1/4 inches long, has 3 long, fleshy petals. Hand pollination—described in connection the atemoya—will increase fruit set, but is usually not required.

FRUIT CHARACTERISTICS The fruit is compound, the exterior divided into a network of bulbous protuberances. The overall shape ranges from globose to conical. The fruit may measure from 2 to 4 inches across, but is typically less than 3 inches in diameter. Skin color may be yellowish-green, bluish-green, grayish-green, pink, or mauve. As the fruit ripens, the exterior segments separate slightly. The flesh is composed of segments of whitish or pinkish flesh, which is sweet, melting, and delicious. Hard, dark brown seeds are present in most segments. While the sugar apple is undeniably a fruit of fine quality, it lacks the complexity of flavor exhibited by superior selections of atemoya.

CULTIVARS Many sugar apples in Florida are grown from seed. A few cultivars have been vegetatively propagated, including 'Kampong Mauve,' 'Purple,' 'Red,' and 'Thai-Lessard.' Although at least 2 seedless varieties exist, they produce fruit of mediocre quality.

RELATIVES Other members of the Annonaceae family profiled within this book include the atemoya, *Annona squamosa* x *Annona cherimola;* the custard apple, *Annona reticulata;* and the pawpaw, *Asimina triloba.* The cherimoya, *Annona cherimola;* ilana, *Annona diversifolia;* soursop, *Annona muricata;* and biriba, *Rollinia deliciosa,* are described in the subsection devoted to relatives of the atemoya.

CLIMATE The sugar apple is tropical in habit, but will survive in subtropical climates. In Florida, it is regularly grown from Palm Beach County and the Ft. Myers area southward. A mature tree may survive a temperature drop to 28° F. The sugar apple is slightly less cold hardy than its cousin, the atemoya.

CULTIVATION The sugar apple has low to moderate maintenance requirements. It will tolerate a wide variety of soils, but cannot withstand poor drainage. It responds well to foliar nutrient sprays. Irrigation during the spring will result in an increase in the size and number of fruit. Irrigation should be suspended during periods of dormancy. The primary pest of the sugar apple in Florida is the chalcid wasp, discussed in connection with the atemoya. Mealybugs can be easily controlled with sprays of insecticidal oil. On occasion, scales will attack the foliage. The sugar apple is frequently grown from seed. Seeds germinate in from 30 to 60 days. Superior clones can be reproduced by shield budding or grafting.

HARVEST AND USE Fruit ripen over the summer and early fall. Each tree yields an average of between 40 and 100 fruit. The fruit is ripe when the segments begin to separate. If left on the tree too long, the fruit will break apart and may be invaded by ants and other insects. The fruit is most often consumed fresh. It can also be used in milkshakes and as a flavoring for ice cream. As with other Annonas, the seeds are toxic and should be removed prior to processing.

The sugar apple is variable in shape, size, and color.

Tamarind

SCIENTIFIC NAME: *Tamarindus indica*
FAMILY: Fabaceae
OTHER COMMON NAME: Tamarindo (Spanish)

Fruiting Calendar

JAN	FEB	MAR	APR	MAY	JUN	JUL	AUG	SEP	OCT	NOV	DEC

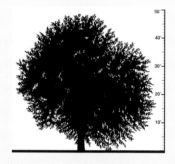

Characteristics

Overall Rating	★★★
Ease of Care	★★★★
Taste/Quality	★★★
Productivity	★★★★
Landscape Value	★★★
Wind Tolerance	★★★★★
Salt Tolerance	★★★★
Drought Tolerance	★★★
Flood Tolerance	★★★
Cold Tolerance	★★

The tamarind was once a common tree in south Florida; widely planted for shade, for its ornamental value, and as a fruit tree. Today, this valuable species is rarely seen outside collections. The fruit, a swollen pod filled with tangy, sour-sweet pulp, is borne in great profusion. The pulp is an important ingredient in Indian cooking and has many uses. The tree is tough, wind resistant, and adaptable, but may become too large for the average suburban lot.

Known Hazards

None

GEOGRAPHIC DISTRIBUTION The tamarind is native to tropical regions of eastern Africa.

TREE DESCRIPTION The tamarind is a large tree that may grow to 80 feet in height. It is often classified as an evergreen, although in Florida it may drop its leaves in early spring or during periods of hot weather. Leaves are alternate, pinnately compound, made up of between 10 and 26 leaflets arranged in pairs. The leaflets are light green, usually less than 1 inch in length. They fold closed during cold weather, after sunset, or when the tree is subjected to stress. The trunk is straight and stout, covered in scabrous, dark-gray bark. The habit of growth is slow to moderate. The tree is long lived and has been known to bear fruit for more than 150 years. Pale, 1-inch flowers are borne on short racemes that emerge from new growth. In Florida, flowering usually occurs in early summer.

FRUIT CHARACTERISTICS The fruit is a brown pod, measuring between 3 and 10 inches in length. The pod swells with maturity. It contains brown pulp, surrounding several flat, brown seeds. A few strands of fiber run through the pulp from the base of the pod to the apex. The pulp is thick, sticky, and pasty, with a distinctive flavor that is sweet, sour, and tangy. As the pod ripens, the pulp shrinks and dehydrates and the skin becomes brittle.

CULTIVARS The tamarind is sometimes grouped into 2 types. East Indian types have long pods containing 6 to 12 seeds. West Indian types have shorter pods containing 3 to 5 seeds. Superior cultivars, when they can be obtained, include 'Markham,' 'Manilla Sweet,' 'Pink,' and 'See Tong.' Several sweet or semi-sweet varieties of tamarind are available in Florida, although these are often reproduced from seed.

RELATIVES The tamarind is a member of the Fabaceae or Legume family (formerly Leguminosae). This immense family, estimated to contain between 7,000 and 19,000 species, contains numerous species of economic import. The carob, *Ceratonia siliqua*, is a medium-size tree native to the eastern Mediterranean region. It has been cultivated for more than 2,000 years. It bears dark pods that can be ground to make a chocolate substitute. The tree is said to be hardy to about 20° F. It has been grown in Florida since the mid-1800s. Superior cultivars include 'Amele,' 'Santa Fe,' and 'Sfax.' Other fruiting legumes include the ice cream bean, *Inga* spp., and the Manila tamarind, *Pithecellobium dulce*.

CLIMATE The tamarind requires a tropical or subtropical climate. Foliage is damaged at about 29° F. The tree may be severely injured or killed when the temperature falls to 25° F.

CULTIVATION The tamarind is one of the least demanding trees described within this book. It will grow on various soil types, including poor soils and oolitic limestone. Like most fruit trees, it requires proper drainage. It is somewhat tolerant of salt spray. It will establish readily in windy sites and can withstand hurricane-force winds. The tamarind should be planted in full sun. The tree is highly drought tolerant and rarely requires irrigation. Periodic applications of a balanced fertilizer will speed growth. Few pests and diseases affect the tree. Scales, mealybugs, aphids, and thrips are sometimes present in small numbers, but rarely require treatment. The tamarind is most often grown from seed. Seeds remain viable for up to a year and germinate readily, often within a week. The tree can also be propagated through grafting and air layering.

HARVEST AND USE Fruit usually ripens during the late fall and early winter. A single tree may produce more than 250 pounds of pods. The pod is harvested by hand. It can be stored on the tree for several weeks after it matures. The pod will store several months at room temperature and may also be refrigerated. The pulp can be used in sauces, gravies, and chutneys. It can be used as a flavoring in drinks and can be combined with sugar to make a delicious candy.

The sticky, sweet, and sour pulp of the tamarind has many uses.

The powerful trunk and root system of the tamarind help it survive hurricane-force winds.

Watermelon

SCIENTIFIC NAME: *Citrullus lanatus*
FAMILY: Cucurbitaceae
OTHER COMMON NAME: Sandía (Spanish)

Fruiting Calendar

JAN	FEB	MAR	APR	MAY	JUN	JUL	AUG	SEP	OCT	NOV	DEC

Characteristics

Overall Rating	★★★
Ease of Care	★★
Taste/Quality	★★★
Productivity	★★★
Landscape Value	★★
Wind Tolerance	★★★
Salt Tolerance	★
Drought Tolerance	★★★
Flood Tolerance	★
Cold Tolerance	★★★

The watermelon is a familiar and much-admired fruit crop. The fruit can attain a very large size, with the record weighing more than 250 pounds. However, over the past few decades, the trend has been toward growing smaller, user-friendly fruit. The watermelon is a sweet, refreshing summertime treat. Its appeal seems to rise with the temperature. It is highly recommended for dooryard planting in all regions of the state.

Known Hazards

None

GEOGRAPHIC DISTRIBUTION The watermelon originated in Africa, possibly in the Kalahari Desert region. It spread around the rim of the Mediterranean at an early date and was widely consumed in Europe by the 1200s. It was introduced into North America prior to the 1630s. Today, Florida is the largest producer within the United States.

PLANT DESCRIPTION The watermelon is an annual vine with an aggressive habit of growth. The leaves are alternate, deeply lobed, and lightly pubescent. The root system is deep and fibrous. An individual vine may produce between 1 and 8 fruit. The watermelon is monoecious: it produces male and female flowers on the same plant. The flowers have 5 pale-yellow petals.

FRUIT CHARACTERISTICS The fruit ranges in shape from round to oblong and cylindrical. The skin color is a shade of green or a combination of shades and can be solid or striped. Flesh color usually ranges from a light pink to deep red. However, some cultivars have orange or yellow flesh. A fruit may contain as many as 1,000 seeds.

CULTIVARS At least 200 watermelon cultivars have been selected. Cultivars are classified as diploid (seeded) or triploid (seedless). Although the triploid plant is sterile, it produces white, rudimentary seeds that are consumed with the flesh. To ensure pollination and fruit set, diploid varieties must be inter-planted with triploid varieties. Triploid cultivars recommended for Florida include 'Sweet Slice,' 'Vanessa,' 'Cooperstown,' 'Genesis,' 'Crimson Trio,' 'Summer Sweet 5244,' 'Tri X Palomar,' and 'Millionaire.' Diploid cultivars recommended for Florida include 'Mardi Gras,' 'Daytona,' 'Regency,' 'Jubalee,' 'Sugar Baby,' 'Sangria,' 'Fiesta,' 'Florida Giant,' and 'Stars N Stripes.' While 'Florida Giant' is capable of producing a very large melon, true giants (melons exceeding 100 pounds) are universally grown from the cultivar 'Carolina Cross.'

RELATIVES The Cucurbitaceae or Gourd family, within the order Cucurbitales, and the subclass Dilleniidae, consists of about 750 species within 50 to 100 genera. It is composed primarily of climbing or creeping vines. Prominent among its members is world's largest fruit, the pumpkin, *Cucurbita maxima,* with the record currently exceeding 1,300 pounds. Other fruiting species include the calabash gourds, *Crescentia* spp.; casabanana, *Sicana odorifera;* the horned melon, *Cucumis metuliferus;* various melons such as the casaba, honeydew, muskmelon, Persian melon, and cantaloupe, *Cucumis melo;* squash, *Cucurbita pepo;* and the wax gourd, *Benincasa hispida.*

CLIMATE The watermelon is a seasonal crop. Florida's warm temperatures and ample sunshine are ideal for producing high-quality fruit.

CULTIVATION The watermelon is moderately demanding in its cultural requirements. It will grow in a wide variety of soils, but must have adequate drainage. Loose or tilled soils have been found to promote root development. The watermelon should be grown in full sun. It requires regular irrigation. However, over-irrigation or heavy rainfall during the later phases of fruit development can lead to bland-tasting fruit. Frequent, light applications of a balanced fertilizer over the course of the growing season will enhance growth and fruit development.

The watermelon is traditionally sewn on raised "hills" or beds, with plants spaced 4 to 6 feet apart. Commercial growers often surround the planting hole with a sheet of black or olive green polyethylene to reduce nematode activity, control weeds, warm the soil, and reduce evaporation. The plant is susceptible to attack by numerous insects, including squash bugs, aphids, larva of the cabbage looper moth, leafminers, thrips, and cucumber beetles. Nematodes also cause considerable damage. Diseases include gummy stem blight, anthracnose, fusarium wilt, fruit blotch, mosaic virus, downy mildew, alternaria, and rind necrosis. The plant is generally grown from seed. Seeds are slow to germinate until the soil temperature reaches 70° F.

HARVEST AND USE Only 1 to 3 fruit should be permitted to form on the vines of most cultivars. Most cultivars grown in Florida ripen from May through July. The fruit is mature when it gives off a soft, hollow sound when tapped and when tendrils near the fruit begin to wilt or turn brown. A watermelon will store for 2 to 3 weeks at room temperature.

Flowers form at leaf nodes along the vine.

Newly formed fruit resemble mature fruit in form and color.

White Sapote

SCIENTIFIC NAME: *Casimiroa edulis*
FAMILY: Rutaceae
OTHER COMMON NAMES: Casimiroa, Zapote Blanco, or
  Matasano (Spanish)

Fruiting Calendar

JAN	FEB	MAR	APR	MAY	JUN	JUL	AUG	SEP	OCT	NOV	DEC

Characteristics

Overall Rating	★★★
Ease of Care	★★★
Taste/Quality	★★★★
Productivity	★★★
Landscape Value	★★
Wind Tolerance	★★★★
Salt Tolerance	★★
Drought Tolerance	★★★★
Flood Tolerance	★★★
Cold Tolerance	★★★

The white sapote is a species with great potential as a dooryard fruit in Florida. Despite its name, it does not belong to the Sapote family, but is a distant citrus relative. The flavor of the custardlike pulp ranges from fair to spectacular. A mature tree is capable of very high yields. The white sapote is a rugged species and will grow and fruit as far north as Orlando.

Known Hazards

The seed is said to be toxic.

GEOGRAPHIC DISTRIBUTION The white sapote is native to the highlands of central Mexico. In California it has been widely planted and improved and is a commercial crop of minor import.

TREE DESCRIPTION The white sapote is a large evergreen tree and can grow to more than 45 feet in height. The canopy is thick and spreading and provides deep shade. Leaves are compound and palmate, typically with 3 to 7 leaflets. Leaflets are bright green, glossy, and lanceolate. They range from 6 to 10 inches in length. The tree has a shallow and aggressive root system. The rate of growth is moderate. Flowers are small, inconspicuous, and greenish-yellow. They are borne on short panicles.

FRUIT CHARACTERISTICS The fruit measures from 2 to 5 inches in diameter. The skin ranges in color from green to golden-yellow. It is thin and tender, but is inedible and bitter. The flesh ranges from white to off-white and is often dotted with minute yellow oil glands. The texture is smooth, moist, and custardlike. The taste is sweet with hints of banana, vanilla, lemon, and peach. The fruit contains 3 to 8 seeds, up to an inch in length, that resemble giant citrus seeds.

CULTIVARS Well regarded cultivars include 'Suebell,' a California cultivar that bears small fruit of superior quality over much of the year; 'Homestead,' a Florida cultivar that is productive, precocious, and readily tolerates Florida's heat and humidity; 'Michele,' a California cultivar that is moderately productive; 'Dade,' a Florida cultivar that bears fruit of excellent flavor that blush yellow at maturity; 'Louise,' a productive cultivar with a lengthy harvest season; 'Wilson,' a productive California cultivar with good cold tolerance; 'McDill,' a California cultivar that produces large, greenish-yellow fruit of excellent flavor; 'Maltby,' a California cultivar that bears very large fruit of good flavor; and 'Smathers,' a woolly-leafed variety that originated in Florida and that consistently bears heavy, high-quality crops under Florida conditions.

RELATIVES The white sapote is a member of Rutaceae family, which includes the *Citrus* genus, and the bael fruit, *Aegle marmelos,* profiled within this book. The woolly-leaf sapote, *Casimiroa tetrameria,* native to southern Mexico and northern Central America, is sometimes classified a subspecies of white sapote.

CLIMATE The white sapote can survive winter temperature drops to 22° or even 20° F. It will grow as far north as St. Augustine on Florida's east coast, and the New Port Richey area on Florida's west coast. The species fruits at least as far south as Homestead.

CULTIVATION The white sapote is a tree of low to moderate maintenance requirements. The regimen of care is similar to that applied to citrus. The tree can produce prodigious amounts of fruit and should not be planted near dwellings, patios, or pools. The tree has exhibited some resistance to hurricane-force winds. It will withstand occasional water logging, but prefers well-drained soil. The young tree tends to be whippy and should be topped at 4 or 5 feet to encourage lateral branching. In Florida, the white sapote has few serious diseases or pests. It will grow readily from seed, but does not come true to seed. Cleft grafting has been employed with a high rate of success.

HARVEST AND USE A mature tree will supply more fruit than the typical homeowner can use. In Florida, most cultivars bear over the summer and fall. A few cultivars produce fruit sporadically over much of the year. A slight color change is often the best indication of maturity. The fruit must be clipped from the branch. If it is pulled from the branch, the area around the stem insert will spoil before the rest of the flesh ripens. The fruit is ready to eat when slightly soft to the touch.

The flower of the white sapote is usually hidden in the foliage of the tree.

Glossary

Aerial root - A root produced above ground. Climbing cacti produce aerial roots.

Aggregate fruit - A fruit composed of mature ovaries from separate pistils of a single flower. The atemoya and blackberry are aggregate fruit.

Air layering - A method of propagation through which roots are forced to develop on the branch of a tree. The rooted branch is then severed and planted.

Alternate leaves - Leaves that occur singly at the leaf nodes, so that 2 leaves do not appear opposite one another along the stem.

Alternate bearing - Describing the tendency of some fruit trees to produce a heavy crop one year followed by a light crop the following year.

Anther - The enlarged tip of a stamen that produces pollen.

Anthracnose - A fungal disease that causes necrotic lesions on the leaves, stems, and fruit of various species.

Apex - The tip, often of a leaf or fruit. In a fruit the apex is located at the end distant from the stem insert.

Areole - A circular cluster of spines on a cactus.

Aril - A fleshy covering around the seed, which develops from the ovule stalk. In some instances, the aril represents the edible portion of the fruit.

Axil - The juncture of the petiole (leaf stalk) and the stem. Flower buds may develop in leaf axils.

Base - The area of a fruit near the stem insert; the area of a leaf near the petiole.

Berry - A fleshy fruit of somewhat homogenous texture that does not contain a stone. Most berries have succulent skin and multiple seeds.

Budding - A propagation technique, similar to grafting, through which a small bud is removed from the selected cultivar and inserted into a notch or cut in the bark of a rootstock.

Bushel - A unit of volume equal to 1/28 of a cubic meter. A bushel of fruit typically weighs about 50 pounds.

Calyx - The lowermost whorl of sepals (modified leaves) within a flower.

Cambium - The soft, thin layer of tissue lying outside the wood and inside the inner bark of trees and shrubs.

Cane - A thin, flexible stem growing directly from the ground or a low branch. Blueberries and brambles produce canes.

Canker - A necrotic lesion in the bark, usually characterized by the death of underlying cambium tissue.

Capsule - A fruit that splits along sutures at maturity. A capsule is derived from a compound ovary, made up of 2 or more carpels.

Carpel - The portion of the pistil of a flower that contains the ovules.

Catkin - A pendulous inflorescence consisting of a single central axis bearing reduced, unisexual flowers.

Central leader system – A pruning system, used primarily on temperate fruit species, whereby the trunk is encouraged to continue vertically through the canopy in the form of a single, strong, central leader. Horizontal branches are permitted to form only at regularly spaced intervals.

Chilling injury - Internal or external injury to the fruit resulting from exposure to low temperature.

Chilling hours - The number of hours, at or below 45° F, required for bud break, flowering, and fruit set in temperate fruit species.

Chlorosis - A condition, caused by nutritional deficiency or disease, where the leaves fail to produce chlorophyll and turn pale yellow or whitish-green.

Compound fruit - A fruit composed of the mature ovaries from separate pistils of multiple flowers.

Compound leaf - A leaf with 2 or more leaflets attached to a single leaf stem or petiole.

Corm - The enlarged, solid, fleshy base of a stem; a vertical underground storage stem.

Corolla - The whorl of petals located above the sepals in a flower.

Cultivar - A cultivated variety; a plant that has been cloned to preserve its genetic characteristics.

Damping-off - A fungal rot that causes the collapse of seedlings shortly after they emerge from the soil.

Deciduous - Describing a plant that sheds its leaves on an annual or seasonal basis.

Dehiscent - Describing a fruit, capsule, or other structure, that splits open upon reaching maturity.

Dioecious - Describing a species that bears male and female flowers on different plants.

Drupe - A fleshy fruit with a stony endocarp containing one or more seeds. Examples include the elderberry, mango, and olive.

Elliptic - Shaped like an ellipse or elongated oval, tapered at both ends.

Embryo - A rudimentary plant present in the seed before germination.

Endocarp - The innermost tissue layer of the pericarp or fruit wall. In stone fruit, the endocarp is the woody pit enclosing the seed.

Entire - Describing a leaf margin with a smooth edge; not containing notches, teeth, or lobes.

Evergreen - Describing a plant that does not lose its leaves on an annual or seasonal basis.

Exocarp - The outermost layer of the pericarp or fruit wall; the rind or skin of the fruit.

Flower - A reproductive structure, ordinarily composed of sepals, petals, stamens, and pistils, although sometimes lacking one or more of these parts.

Fruit - The mature ovary of a flowering plant, enclosing the seeds.

Glabrous - Describing a smooth, hairless surface.

Globose - Round or rounded.

Graft union - The point at which rootstock and scion are united.

Grafting - A means of propagation through which a branch tip from one plant (the scion) is bonded to the stem and root system of a second plant (the rootstock).

Herb - A short-lived plant with pliable, nonwoody stems.

Hesperidium - A specialized berry with a leathery rind and a segmented interior.

Heterogamous - Describing a plant that produces both male and female flowers.

Homogamous - Describing a plant that produces flowers of only one sex.

Indehiscent - Describing a fruit or nut which does not split open at maturity to discharge its seeds.

Inflorescence - A cluster of flowers.

Invasive exotic - A nonnative plant that has escaped cultivation and that is reproducing and spreading.

Lanceolate - Describing a narrow, lance-shaped leaf, broadest at the base and tapered to a long point.

Leaf - A plant organ consisting of a blade and a petiole.

Leaflet - The individual blade of a compound leaf.

Lobed - Describing a leaf with extensions separated by deep indentations, known as sinuses.

Margin - The edge of a leaf.

Mesocarp - The middle layer of the pericarp or fruit wall.

Monoecious - Bearing separate male and female flowers on the same plant.

Multiple fruit - Fruit produced by the fusion or adherence of 2 or more ovaries arising from different flowers. The pineapple is a multiple fruit.

Nematodes - Microscopic worms living in the soil, some of which attack plant roots.

Node - The area of a stem where the vascular tissue branches into leaves, buds, or other appendages.

Nut - A dry, indehiscent fruit with a hard pericarp.

Oblong - Tapered at both ends, but with the sides nearly parallel.

Oblate - Oval flattened vertically.

Obovate - Oval with the broadest part near the apex.

Opposite leaves - Leaves arranged so that 2 occur at each node.

Ovary - The part of the flower that encloses the ovules; the enlarged base of the pistil, which becomes the fruit.

Ovate - Oval, with the broader part near the base.

Ovule - The structure within a flowering plant that gives rise to the seed.

Palmate - Describing leaflets, leaf structures, or leaf veins, arising from and spreading outward from a single point at the base of the leaf.

Pedicel - The flower stalk.

Peduncle - The main stalk of an inflorescence.

Perfect flower - A perfect flower is hermaphroditic and contains female parts (pistils) and male parts (stamens).

Pericarp - The fruit wall, composed of the endocarp, the mesocarp, and the exocarp.

Persistent - Remaining attached.

Petal - A unit of the corolla, a flower appendage located between the sepals and the stamens.

Petiole - The leaf stalk.

Phloem - Stem or trunk tissue involved in the transport of the products of photosynthesis.

Pinnately compound - Containing leaflets arranged oppositely or alternately along the leaf stalk axis.

Pinnately veined - Describing leaves with a middle vein, intersected by secondary veins, which radiate toward the margins on each side.

Pistil - The female part of a flower, comprising the central organs, and composed of an ovary, style, and stigma.

Pod - A dehiscent, often dry, fruit that develops from a single carpel that splits along 2 parallel sutures.

Pollen - The male microspore of seed plants, produced by the stamens.

Pollination - The process of transferring the pollen from the stamen to the stigma.

Polygamous - Producing unisexual and bisexual flowers on the same plant.

Pome - A fruit with a papery or cartilaginous endocarp (the core) enclosed within a swollen, fleshy receptacle.

Precocious - Coming into bearing at a young age.

Protogyrus - Producing flowers that change sex.

Pseudocarp - A false fruit that develops, at least in part, from nonovarian tissue.

Pyriform - Pear-shaped.

Pubescent - Having fine hairs on the surface.

Raceme - A long, often pendulant, unbranched inflorescence.

Receptacle - The terminal part of the pedicel or flower stalk to which the flower parts are attached.

Reniform - Describing a rounded leaf shape with a broad base; like the leaf of a seagrape.

Rhizome - A horizontal underground stem, such as is found in bananas.

Rootstock - The root system and lower stem of a grafted tree; all portions below the graft union.

Scion - A cultivar or selected plant grafted on a rootstock.

Seed - A fertilized, matured ovule; a multicellular reproductive structure containing the plant embryo.

Self-compatible - Describing a plant which does not require cross-pollination in order to produce fruit.

Sepal - Part of the calyx; the outermost structure of a flower.

Serrate - Describing a leaf margin with sharp, forward-pointing teeth.

Simple fruit - A fruit that develops from the ovary of a solitary pistil in a single flower.

Simple leaf - Any leaf that is not compound.

Spathe - A sheathlike leaf partially enclosing an inflorescence.

Stamen - The male pollen-producing organ of a flower, composed of a filament and anther.

Stigma - The sticky tip of the pistil of a flower receptive to pollen.

Stone fruit - A drupe containing a single seed surrounded by a hard shell and covered by pulp.

Strain - A sub-subspecies of a plant, or a variant of a cultivar that is nearly identical, not deserving full cultivar status.

Style - Part of the pistil of a flower. The style is a tubular structure that connects the stigma and the ovary.

Sucker - A vigorous shoot sprouting from below ground or from older growth.

Taproot - A well-developed, vertical, primary root.

Thinning - The process of removing flowers or immature fruit to increase the ultimate size of the remaining fruit.

Top working - A method of propagation that involves replacing the entire canopy of an established tree by cutting the tree back severely and by grafting a new cultivar to shoots that arise from the stump.

Toothed - Toothed or dentate leaves have small teeth along the margins.

Vase system - A pruning system that encourages formation of a short trunk ending in 3 to 5 main branches, which extend upward and outward to form a vase shape. The center of the canopy is open and inward-pointing branches are periodically removed.

Venation - Referring to the pattern of veins within the blade of a leaf. Venation may be palmate, pinnate, or parallel.

Whorl - A group of leaves, petals, or carpels forming a spokelike pattern around a central axis.

Xylem - Trunk and stem tissue involved in the transport of water and minerals upward to the canopy.

Yield - The amount of fruit produced by a tree.

Organizations

Rare Fruit Council International, Inc.
P. O. Box 660506
Miami Springs, FL 33266

California Rare Fruit Growers, Inc.
The Fullerton Arboretum - CSUF
P.O. Box 6850
Fullerton, CA 92834–6850

North American Fruit Explorers
P.O. Box 94
Chaplin, IL 62628

Collections and Arboretums

FRUIT AND SPICE PARK Miami-Dade County's Fruit and Spice Park houses the premier public collection of subtropical and tropical fruit trees in the continental United States. This unique park, set on 33 acres, was founded in 1944. It is located in the Redlands area, at 24801 SW 187 Avenue, Homestead, FL 33031. The park features mature specimens of many rare and exotic fruit species. Visitors can stroll beneath mature jackfruit, star apple, lychee, and tamarind trees, among others. The park has an extensive collection of mangos, avocados, and Annonas. Samples are available for tasting at the front entrance. While visitors are prohibited from picking fruit, they are permitted to taste fallen fruit.

FAIRCHILD TROPICAL BOTANIC GARDEN Fairchild Tropical Garden has an extensive collection of fruit trees, palms, and ornamentals from around the world. This lush 83-acre arboretum was founded in 1938. It is located just south of downtown Miami, at 10901 Old Cutler Road, Coral Gables, FL 33156. Fairchild Tropical Garden hosts the International Mango Festival and fruit seminars and other events. The Whitman Tropical Fruit Pavilion features a number of ultra-tropical fruit trees in a specially constructed greenhouse. Other fruiting species are scattered throughout the garden.

MOUNTS BOTANICAL GARDEN Mounts Botanical Garden is located in West Palm Beach, just west of the airport. The address 531 North Military Trail, West Palm Beach, FL 33415. It features a number of tropical and subtropical fruit trees.

FLAMINGO GARDENS This park features a collection of citrus and some other subtropical fruit species. It is located in Broward County, west of Ft. Lauderdale, at 3750 S. Flamingo Road, Davie, FL 33330. Tours and educational programs are available.

HARRY P. LEU GARDENS While this botanical garden is primarily a collection of flowering plants, it also features several tropical and subtropical fruit species. It is located at 1920 N. Forest Avenue, Orlando, FL 32803.

AUDUBON BOTANICAL GARDEN The Botanical Garden at the Tropical Audubon Society is a 3-acre garden located behind the Doc Thomas house in south Miami. It is home to several native fruiting plants. The address is 5530 Sunset Drive, Miami, FL 33143.

EDUCATIONAL CONCERNS FOR HUNGER ORGANIZATION (ECHO) This nonprofit Christian organization, which provides agricultural know-how to those living in impoverished conditions, maintains a demonstration farm, tropical fruit orchard, bookstore, and nursery at 17391 Durrance Road, North Fort Myers, FL 33917.

MARIE SELBY BOTANICAL GARDENS This garden promotes conservation of tropical plants. It is home to more than 20,000 plants from 214 families, including several fruiting species. Marie Selby Botanical Gardens is located at 811 South Palm Avenue, Sarasota, FL 34236.

Nurseries

Chestnut Hill Tree Farm
15105 NW 94th Avenue
Alachua, FL 32615
386-462-2820
Located just north of Gainesville, Chestnut Hill has an extensive collection of low-chill temperate fruit trees and other transitional species.

Excalibur Fruit Trees
5200 Fearnley Road
Lake Worth, FL 33467
561-969-6988
Richard and Lynda Wilson's nursery has long been *the* destination for those seeking the rare and exotic. It is home to one of the largest and most diverse collections of tropical and subtropical fruit plants anywhere.

Garden of Delights Nursery
14560 SW 14th Street
Davie, FL 33325
954-370-9004

Going Bananas
24401 SW 197 Avenue
Homestead, Florida 33031
305-247-0397

Hopkins Tropical Fruit Nursery
25355 Shultz Grade
Immokalee, FL 34142
239-658-0370

Jene's Tropicals
6831 Central Avenue
St. Petersburg, FL 33710
727-344-1668

Josan Growers
5418 Fearnley Road
Lake Worth, FL 33467
561-968-2466
This nursery carries large container-grown tropical and subtropical fruit trees. Its jackfruit collection is particularly impressive.

Just Fruits and Exotics
30 St. Francis Street
Crawfordville, FL 32327
850-926-5644
Located 19 miles south of Tallahassee, this nursery is an excellent source for hard-to-find fruit trees suitable for planting in north Florida. Staff is knowledgeable and helpful. This nursery has an impressive collection of mature citrus, which is protected from winter cold by frame enclosures.

Lychee Tree Nursery
3151 South Kanner Highway
Stuart, FL 34994
772-283-4054

Lara Farms Nursery
18660 SW 200 Street
Miami, FL 33187
305-253-2750
This nursery specializes in mamey sapote.

Our Kids Tropicals Nursery
17229 Phil Peters Road
Windermere, FL 32787
407-877-6883

Pine Island Nursery
16300 SW 184th Street
Miami, FL 33187
305-233-5501
This is the premier fruit tree nursery in the Miami area. Pine Island carries a huge assortment of tropical and subtropical species.

Plant Creations
28301 SW 172nd Avenue
Homestead, FL 33030
305-248-8147
Plant Creations carries several difficult-to-locate native fruit trees.

Ray's Nursery
18905 SW 177th Avenue
Miami, FL 33187
305-255-3589

Richard Lyons Nursery
20200 SW 134th Avenue
Miami, FL 33125
305-251-6293

Rockledge Gardens
2153 South US Highway #1
Rockledge, Florida 32955
321-636-7662

The Treehouse Nursery
Bokeelia, FL 33922
239-283-3688
This nursery, located west of Ft. Myers on Pine Island, has a vast selection of tropical and subtropical fruit trees, including varieties that cannot be found elsewhere.

Truly Tropical Fruit Trees
2750 Seacrest Boulevard
Delray Beach, FL 33444
561-278-7754

Zone Ten Nursery
18900 S.W. 186 Street
Miami, FL 22187
305-255-9825

Native Fruiting Plants

This list presents Florida's native fruiting plants, roughly ordered from top to bottom based upon their overall value as fruiting plants.

Note: some of the species listed below are endangered, require preparation prior to consumption, or produce adverse reactions if consumed in quantity. Extreme caution should be exercised in the collection and sampling of any native fruit.

COMMON NAME	SCIENTIFIC NAME
Muscadine grape	*Vitis rotundifolia*
Red mulberry	*Morus rubra*
Pawpaw	*Asimina* spp.
Blackberry	*Rubus* spp.
American persimmon	*Diospyros virginiana*
Black walnut	*Juglans nigra*
Blueberry	*Vaccinium* spp.
Maypop	*Passiflora incarnata*
Mayhaw	*Crataegus* spp.
Seagrape	*Coccoloba uvifera*
Groundcherry	*Physalis* spp.
Chinkapin	*Castanea* spp.
Beech	*Fagus grandifolia*
Stoppers	*Myrcianthes fragrans, Eugenia* spp.
Chickasaw plum	*Prunus angustifolia*
Prickly pear	*Opuntia* spp.
Summer haw	*Crataegus* spp.
Wild grape	*Vitis* spp.
Ogeechee lime	*Nyssa ogeche*
Hickory	*Carya* spp.
Mayapple	*Podophyllum peltatum*
Seven-year apple	*Casasia clusiifolia*
Mastic	*Sideroxylon foetidissimum*
Huckleberry	*Gaylussacia* spp.
Cocoplum	*Chrysobalanus icaco*
Miccosukee gooseberry	*Ribes echinellum*
Pricklyapple cactus	*Harrisia* spp.
Pond apple	*Annona glabra*
Black cherry	*Prunus serotina*
American plum	*Prunus americana*
Elderberry	*Sambucus nigra*
Hackberry	*Celtis* spp.
Pignut hickory	*Carya glabra*
Florida hickory	*Carya floridana*
Oaks	*Quercus* spp.
Serviceberry	*Amelanchier* spp.
Wild dilly	*Manilkara bahamensis*
Tallowwood	*Ximenia Americana*
Safron plum	*Sideroxylon celastrinum*
Satinleaf	*Chrysophyllum oliviforme*
Southern crabapple	*Malus anggustifolia*
Beautyberry	*Callicarpa americana*
Barberry	*Berberis bealei*

References and Further Reading

Bermejo, J. E. Hernándo, and J. León. *Neglected Crops: 1492 from a Different Perspective*. Food and Agriculture Organization of the United Nations, 1994.

Campbell, Richard, and Noris Ledsema. *The Exotic Jackfruit: Growing the World's Largest Fruit*. Miami, FL: Fairchild Tropical Garden, 2003.

Campbell, Richard, Carl Campbell, and Noris Ledsema. *Tropical Mangos: How to Grow the World's Most Delicious Fruit*. Miami, FL: Fairchild Tropical Garden, 2002.

Campbell, Richard, ed. *Mangos: A Guide to Mangos in Florida*. Miami, FL: Fairchild Tropical Garden, 1992.

Erez, Amnon. *Temperate Fruit Crops in Warm Climates* Boston, MA: Kluwer Academic Publications, 2000.

Facciola, Stephen. *Conucopia II: A Source Book of Edible Plants*. Vista, CA: Kampong Publications, 1998.

Fact Sheets: CIR1034, CIR1192, CIR1271, CIR1400, CIR1440, EN388, ENH334, ENH368, ENH390, ENH408, ENH563, ENH564, ENH672, ENH687, ENH776, ENH836, FC6, FC9, FC9, FC28, FC30, FC47, FCS8516, FCS8522, FCS8537, FS40, HS1, HS2, HS4-HS7, HS10-HS12, HS14, HS16A, HS23, HS27, HS37-HS39, HS41, HS44, HS49, HS50, HS60, HS63, HS64, HS169, HS171, HS176, HS763, HS764, HS807, HS882, HS920, HS967, HS984, PP21, SP101, Gainesville, FL: Florida Cooperative Extension Service, Institute of Food and Agricultural Sciences, University of Florida (UF/IFAS).

Fairchild, David. *The World Was My Garden*. New York, NY: Charles Scribner's Sons, 1944.

Fruit Gardener, Vols. 22–37. California Rare Fruit Growers, 1990–2005.

Fruit Facts (http://www.crfg.org/pubs/frtfacts.html). California Rare Fruit Growers, 1996–2001.

Glowinski, Louis. *The Complete Book of Fruit Growing in Australia*. Port Melborne, Victoria, Australia: Lothian Books, 1991.

Hessayon, D. G. *The Fruit Expert*. London, UK: Expert Books, 1990.

Jackson, Larry K., and Frederick S. Davies. *Citrus Growing in Florida*. Gainesville, FL: University Press of Florida, 1999.

Jensen, Michael. *Trees Commonly Cultivated in Southeast Asia: An Illustrated Field Guide*. Bangkok, Thailand Food and Agricultural Organization of the United Nations, Regional Office for Asia and the Pacific, 1999.

Krezdorn, A. H., Lewis S. Maxwell, Eric V. Golby, and Albert A. Will. *Florida Fruit: Fresh Fruit for the Home How to Choose and Grow Them*. Tampa, FL: Lewis S. Maxwell, 1967.

Livingston, Bruce *Rules of Thumb*, article.

Manuel, Leo, ed. *Rare Fruit News Online* (biweekly newsletter http://www.rarefruit.com), 1996–2006.

Morton, Julia. *Fruits of Warm Climates*. Miami, FL: Julia Morton, 1987.

Morton, Julia. *Wild Plants for Survival in South Florida*. Miami, FL: Fairchild Tropical Garden, 1982.

Nakasone, Henry Y., and Robert E. Paull. *Tropical Fruits*. New York, NY: CABI Publishing, 1998.

Nelson, Gil. *Florida's Best Native Landscape Plants, 200 Readily Available Species for Homeowners and Professionals*. Gainesville, FL: University Press of Florida, 2003.

Nelson, Gil. *The Trees of Florida: A Reference and Field Guide*. Sarasota, FL: Pineapple Press, 1994.

Ortho Books, *All About Citrus & Subtropical Fruits*. San Ramon, CA: Ortho, 1985.

Otto, Stella. *The Backyard Berry Book: A Hands-On Guide to Growing Berries, Brambles, and Vine Fruit in the Home Garden*. Maple City, MI: Ottographics, 1995.

Otto, Stella. *The Backyard Orchardist: A Complete Guide to Growing Fruit Trees in the Home Garden*. Maple City, MI: Ottographics, rev. ed. 1995.

Popenoe, Wilson. *Manual of Tropical and Subtropical Fruits, Excluding the Banana, Coconut, Pineapple, Citrus Fruits, Olive, and Fig*. New York, NY: MacMillan, 1924.

Reich, Lee. *Uncommon Fruits for Every Garden*. Portland, OR: Timber Press, 2004.

Reich, Lee. *Uncommon Fruits Worthy of Attention*. New York, NY: Addison-Wesley, 1991.

Rosenblum, Mort, *Olives: The Life and Lore of a Noble Fruit*. New York, NY: North Point Press, 1996.

Rushing, Felder, and Walter Reeves. *Herbs, Fruits and Vegetables For Georgia: 50 Great Plants For Georgia Gardens*. Nashville, TN: Cool Springs Press, 2003.

Silva, Silvestre, and Helena Tassara. *Fruit in Brazil.* Sao Paulo, Brazil: Empressa das Artes, 1996.

Stresau, Frederic B. *Florida My Eden.* Port Salerno, FL: Florida Classics Library, 1986.

Tate, Desmond. *Tropical Fruit.* Singapore, Malaysia: Butterworth-Heinemann, 1999.

Tropical Fruit News (Vols. 1–40). Miami Springs, FL: Rare Fruit Council International, 1967–2006.

Van Aken, Norman. *The Great Exotic Fruit Book: A Handbook of Tropical and Subtropical Fruits, With Recipes.* Berkeley, CA: Ten Speed Press, 1995.

Van Atta, Marian. *Exotic Foods: A Kitchen and Garden Guide.* Sarasota, FL: Pineapple Press, 2002.

Vietmeyer, Noel D., ed. *Lost Crops of the Incas: Little Known Plants of the Andes with Promise for Worldwide Cultivation.* Washington, DC: National Academy Press, 1989.

Walheim, Lance. *Citrus: Complete Guide to Selecting & Growing More Than 100 Varieties for California, Arizona, Texas, the Gulf Coast and Florida.* Tuscon, AZ: Ironwood Press, 1996.

Whiley, A.W., B. Schaffer, and B. N. Wolstenholme, eds. *The Avocado: Botony, Production and Uses.* New York, NY: CABI Publishing, 2002.

Whitman, William F. *Five Decades with Tropical Fruit: A Personal Journey.* Miami, FL: Fairchild Tropical Garden, 2001.

White, Hazel, and Janet H. Sanchez. *The Edible Garden.* Menlo Park, CA: Sunset, 2005.

Index

Illustrations and photographs indicated with **boldface** text.

Here are some other books from Pineapple Press on related topics. For a complete catalog, write to Pineapple Press, P.O. Box 3889, Sarasota, Florida 34230-3889, or call (800) 746-3275. Or visit our website at www.pineapplepress.com.

Groundcovers for the South by Marie Harrison. Presents a variety of plants that can serve as groundcovers in the American South. Each entry gives detailed information on ideal growing conditions, plant care, and different selections within each species. Color photographs and line drawings make identification easy. (pb)

Southern Gardening: An Environmentally Sensitive Approach by Marie Harrison. A comprehensive guide to beautiful, environmentally conscious yards and gardens. Suggests useful groundcovers and easy-care, adaptable trees, shrubs, perennials, and annuals. (pb)

Gardening in the Coastal South by Marie Harrison. A Master Gardener discusses coastal gardening considerations such as salt tolerance; environmental issues such as pesticide use, beneficial insects, and exotic invasives; and specific issues such as gardening for butterflies and birds. Color photos and charming pen-and-ink illustrations round out the text. (pb)

Flowering Trees of Florida by Mark Stebbins. Written for both the seasoned arborist and the weekend gardener alike, this comprehensive guide offers 74 outstanding tropical flowering trees that will grow in Florida's subtropical climate. Full-color photos throughout. (pb)

Landscaping in Florida by Mac Perry. A photo idea book packed with irresistible ideas for inviting entryways, patios, pools, walkways, and more. Over 200 photos and eight pages of color photos. (pb)

Ornamental Tropical Shrubs by Amanda Jarrett. Stunning color photos and full information profile for 83 shrubs. (hb & pb)

The Trees of Florida by Gil Nelson. The first comprehensive guide to Florida's amazing variety of tree species, this book serves as both a reference and a field guide. (hb & pb)